STRESS PHYSIOLOGY AND FOREST PRODUCTIVITY

FORESTRY SCIENCES

Baas P, ed: New Perspectives in Wood Anatomy. 1982. ISBN 90-247-2526-7
Prins CFL, ed: Production, Marketing and Use of Finger-Jointed Sawnwood. 1982.
 ISBN 90-247-2569-0
Oldeman RAA, et al., eds: Tropical Hardwood Utilization: Practice and Prospects. 1982.
 ISBN 90-247-2581-X
Den Ouden P and Boom BK: Manual of Cultivated Conifers: Hardy in Cold and Warm-
 Temperate Zone. 1982. ISBN 90-247-2148-2
Bonga JM and Durzan DJ, eds: Tissue Culture in Forestry. 1982. ISBN 90-247-2660-3
Satoo T and Magwick HAI: Forest Biomass. 1982. ISBN 90-247-2710-3
Van Nao T, ed: Forest Fire Prevention and Control. 1982. ISBN 90-247-3050-3
Douglas J: A Re-appraisal of Forestry Development in Developing Countries. 1983.
 ISBN 90-247-2830-4
Gordon JC and Wheeler CT, eds: Biological Nitrogen Fixation in Forest Ecosystems:
 Foundations and Applications. 1983. ISBN 90-247-2849-5
Németh MV: The Virus-Mycoplasma and Rikettsia Disease of Fruit Trees.
 ISBN 90-247-2868-1
Duryea ML and Landis TD, eds: Forest Nursery Manual: Production of Bareroot Seed-
 lings. 1984. ISBN 90-247-2913-0
Hummel FC, ed: Forest Policy: A Contribution to Resource Development. 1984.
 ISBN 90-247-2883-5
Manion PD, ed: Scleroderris Canker of Conifers. 1984. ISBN 90-247-2912-2
Duryea ML and Brown GN, eds: Seedling Physiology and Reforestation Success. 1984.
 ISBN 90-247-2949-1
Staaf KAG and Wiksten NA: Tree Harvesting Techniques. 1984. ISBN 90-247-2994-7
Boyd JD: Biophysical Control of Microfibril Orientation in Plant Cell Walls. 1985.
 ISBN 90-247-3101-1
Findlay WPK, ed: Preservation of Timber in the Tropics. 1985. ISBN 90-247-3112-7
Samset I: Winch and Cable Systems. 1985. ISBN 90-247-3205-0
Leary RA: Interaction Theory in Forest Ecology and Management. 1985.
 ISBN 90-247-3220-4
Gessel SP: Forest Site and Productivity. 1986. ISBN 90-247-3284-0
Hennessey TC, Dougherty PM, Kossuth SV and Johnson JD, eds: Stress Physiology and
 Forest Productivity. 1986. ISBN 90-247-3359-6

Stress physiology and forest productivity

Proceedings of the Physiology Working Group Technical Session. Society of American Foresters National Convention, Fort Collins, Colorado, USA, July 28–31, 1985

edited by

THOMAS C. HENNESSEY
Oklahoma State University, Stillwater, Oklahoma, USA

PHILLIP M. DOUGHERTY
University of Georgia, Athens, Georgia, USA

SUSAN V. KOSSUTH
USDA Forest Service, Gainesville, Florida, USA

JON D. JOHNSON
University of Florida, Gainesville, Florida, USA

1986 **MARTINUS NIJHOFF PUBLISHERS**
a member of the KLUWER ACADEMIC PUBLISHERS GROUP
DORDRECHT / BOSTON / LANCASTER

Distributors

for the United States and Canada: Kluwer Academic Publishers, 190 Old Derby
Street, Hingham, MA 02043, USA
for the UK and Ireland: Kluwer Academic Publishers, MTP Press Limited,
Falcon House, Queen Square, Lancaster LA1 1RN, UK
for all other countries: Kluwer Academic Publishers Group, Distribution Center,
P.O. Box 322, 3300 AH Dordrecht, The Netherlands

Library of Congress Cataloging in Publication Data

```
Stress physiology and forest productivity.

   (Forestry sciences ; v.    )
   Bibliography: p.
   1. Trees--Physiology--Congresses.  2. Plants, Effect
of stress on--Congresses.  3. Forest ecology--Congresses.
4. Forest management--Congresses.  5. Forest productivity
--Congresses.  I. Hennessey, Thomas C.  II. Society of
American Foresters.  Physiology Working Group.
III. Society of American Foresters.  Convention (1985 :
Fort Collins, Colo.)
SD395.S77  1986        634.9'6           86-8644
```

ISBN-13: 978-94-010-8469-7   e-ISBN-13: 978-94-009-4424-4
DOI: 10.1007/978-94-009-4424-4

SAF 86-04

Copyright

CONTENTS

PREFACE

Maintaining or increasing stand productivity is the concern of forest land managers worldwide. Consequently, there is increasing interest in understanding the impact of environmental stress on productivity and the development of management strategies that ameliorate or reduce the deleterious effects.

Invited scientists gathered in Fort Collins, Colorado on July 30, 1985, to present the current state of knowledge regarding the impact of environmental stress on forest stand productivity. Particular attention was given to elucidating the mode of action by which individual stress elements reduce productivity. Environmental factors and the levels that constitute stressed (suboptimal) conditions in forest stands were identified, and the effects of stress intensity and duration on key stand parameters, including photosynthesis, respiration, assimilate partitioning, senescence and mortality, were emphasized.

The role of genetics and silvicultural treatments in lessening the stress impact on stand productivity was presented, particularly in regards to alternative methods for environmental stress management. Modeling of stand dynamics in response to environmental stress was explored as an effective research and management tool.

Improved forest management practices will develop as we improve our understanding of the nature of important environmental stresses and as we comprehend their impact on tree and stand performance, manifested through physiological processes and genetic potential. This book is dedicated to such an understanding and comprehension.

Thomas C. Hennessey
Phillip M. Dougherty
Susan V. Kossuth
Jon D. Johnson

1. INTRODUCTORY OVERVIEW

S. G. PALLARDY

Associate Professor, School of Forestry, Fisheries, and Wildlife, University of Missouri, Columbia, Missouri 65211

ABSTRACT

Environmental stresses, both biotic and abiotic, commonly reduce plant growth below that which would be indicated by genetic potential. Physiological processes of plants integrate environment and heredity to control growth. The responses of physiological processes to environmental factors are complex, but progress is being made in our understanding of these responses and in the ability to predict growth from knowledge of environmental modulation of plant physiology. The most successful physiologically-oriented growth models have employed empirically determined relationships between dominant environmental factors and stand growth responses. The focus of this technical session will center on research directed toward improving management practices and growth modeling through a better understanding of fundamental physiological relationships among environment, hereditary potential of forest trees, and stand growth.

1.1 INTRODUCTION

The environment presented to forest stands, artificially or naturally regenerated, managed or unmanaged, commonly reduces growth and wood quality below the maximum attainable level. This fact is not obvious because these limitations are so pervasive that we have relatively few

Contribution of the Missouri Agricultural Experiment Station, Journal Series No. 9970

examples of unrestricted growth with which to compare the
norm. In this technical session, the nature of important
environmental limitations, how they influence stand performance through effects on individual trees, and contemporary
research aimed at increasing forest stand growth and quality
by modifying both plant and environment will be discussed.

1.2 STRESS CONCEPTS

Although most people have an intuitive feeling for the
concept of stress in biology, exact definitions are elusive.
The majority of workers in stress research accept the
concepts, if not terminology, advanced by Levitt (6). In
Levitt's scheme, in which an analogy between biological and
physical stresses is drawn, stress is defined as "any environmental factor capable of inducing a potentially injurious
strain in living organisms." The strain(s) induced are
physical and chemical responses within the organism that
eventually produce the suite of macroresponses we observe as
growth reduction, injury, or death. Continuing Levitt's
mechanical analogy, organisms generally exhibit a capacity
for elastic strain, where physiological responses are only
temporarily altered, but return to normal after the stress
is relieved, followed by plastic strain under greater levels
of stress, where permanent alteration (and often injury and
death) follow severe stress imposition and relief. The
point at which elastic strain turns to plastic strain is
called the "yield point" and it is important because it
marks the level of stress where severe and irreversible
effects begin. The yield point is not static but may change
with moderate stresses of longer duration, as is exemplified
by the process of osmotic adjustment (e.g., 8).

The use of the term "strain" for the physiological responses of organisms to environmental factors has not been
widely adopted; if so, we would be talking about "strain"
management instead of stress management today, and as Paul
Kramer has pointed out (4), no one uses or is likely to use
the term "plant water strain." In practice, stress is used

in a broad sense, describing both deleterious levels of environmental factors and the condition of the plant. Considerations of stress also include the influence of biotic as well as abiotic factors in the plant environment. Levitt's ideas do, however, provide a framework through which one can understand the response of forests to environmental influences. The forest scientist's and manager's jobs are to understand the nature of these elastic and plastic responses and to work toward maintaining stands under minimal "elastic strain."

1.3 ENVIRONMENTAL AND GENETIC INFLUENCES ON STAND GROWTH

The main objective of commodity forestry is to derive raw materials by manipulating the natural growth process through various management practices. In relating this session to practical considerations, two questions are particularly relevant: 1) What pattern of influence does the environment (broadly defined) have on growth? and 2) How are these environmental effects translated into observed changes in growth? Justus von Liebig, arguably the original agricultural chemist, first addressed the former question when he formulated his Law of the Minimum: "The growth of a plant is dependent upon the amount of foodstuff [i.e., nutrients, water, etc.] that is presented to it in minimum quantities." A generalized response to an environmental factor is characterized by a region of direct response to the factor followed by a region of satiation or saturation (9). In the early part of this century Blackman (1) developed this idea further, demonstrating that the saturation level of a yield factor can be increased by the addition of a second limiting factor. These ideas provide a conceptual picture of the influence of environmental factors (particularly nutrition and light) on stand growth processes, but they by no means cover all situations. For example, growth responses to water availability would present an optimum response near field capacity, with reduced growth at lower soil water contents as water stress increases or at higher water contents where flooding stresses are induced. Likewise, winter injury in temperate deciduous species (see Chapter 7) does not follow this saturation response with respect to minimum winter temperatures. Rather, xylem parenchyma mortality is observed across a very narrow band of temperature and is associated with a specific physical event. Hence we should expect that the pattern of influence of individual environmental factors on growth to be diverse.

With regard to the second question, Klebs (2, 3) and Lundegardh (7) have noted that growth is the result of the interaction of environmental factors with the genetic potential of organisms, as mediated by physiological processes (Figure 1). For example, a deficiency of soil phosphorus results in reduced levels of key phosphorus-containing compounds within the plant, a condition which reduces growth below the genetic maximum. In this situation, phosphorus fertilization may bring growth closer to the genetic potential of the tree. Alternatively, tree improvement efforts may produce a new, phosphorus-efficient genotype which, in interacting with its unamended environment, conducts its physiology in such a way as to increase growth over other genotypes. Hence environmental

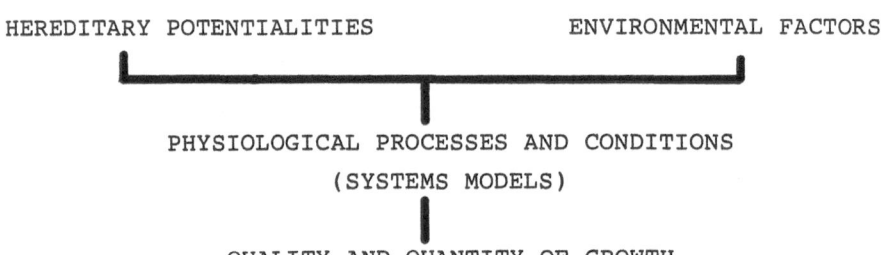

"KLEBS'S CONCEPT"

HEREDITARY POTENTIALITIES   ENVIRONMENTAL FACTORS

PHYSIOLOGICAL PROCESSES AND CONDITIONS
(SYSTEMS MODELS)

QUALITY AND QUANTITY OF GROWTH

FIGURE 1. Representation of the manner in which heredity and environment interact to influence the quantity and quality of growth. Systems models are mathematical representations of such environmental control over physiology. After (5).

manipulation and genetic alteration offer two opportunities to control stand growth. Historically, intuition and empirical experimentation have permitted environmental manipulation to some extent, but in the last generation forest scientists have begun to move toward better understanding of how a tree's environment influences physiological processes. This will be an area of focus today.

Genetic improvement has been slow because of the innate problems encountered in dealing with plants with long periods of juvenility, but tree breeding programs, in several regions of the country and for several major timber species, have progressed to the stage where benefits are accruing. Novel techniques for genetic manipulation (i. e., "genetic engineering") offer potential for tree improvement at a more rapid pace, but substantial work remains before these methods can be exploited in forest tree improvement.

1.4 PHYSIOLOGICALLY-ORIENTED MODELS OF TREE GROWTH

Of interest to many forest managers and researchers is prediction-oriented modeling of plant processes such as physiological response to environment and, ultimately, growth (tools of obvious economic and management importance). Conceptually, physiological growth models can easily be related to Klebs's concepts. Relationships between genetic potential and physiological responses to environment and yield must be developed. In the ultimate model, the physiology of the organism is completely reduced to descriptive equations (mechanistic or empirical) that account for responses of a particular genotype to environment and all its interactions. However, we are far from possessing this type of model, and it is doubtful whether such complete models are within our reach given our present understanding of tree physiology, or whether they are even needed for modeling to produce useful economic and research results. To be certain, a definitive model (or even a close approach) would be a marvelous research tool, but it is unlikely to appear in the near future. Rather, to date most physiological modeling attempts

in forestry have been wisely limited to the "art of the soluble," where productivity is modeled in a situation in which
dominant driving variables are operating (e.g., light, moisture, nutrients). These models have produced useful results.

1.5 EMPHASES OF THE TECHNICAL SESSION

The particular stress that is limiting to tree growth
varies widely during the life cycle of a species. In the
seedling stage light stress is often a dominant factor,
while in fully-occupied, pole-sized stands water and nutrient stresses are often critical. As stands become overmature, disease and insect stresses become especially apparent. However, these stresses may be important at any stage
under certain conditions. In this session, most attention
will be directed toward stands in the intermediate stage of
growth between youth and senescence.

A number of common themes will become apparent in this
session. First, it will be quite clear that environmental
and biological stresses are a fact of existence for all
forest stands, significantly reducing growth below that
which would be theoretically and economically obtainable
with altered management. One valuable contribution of physiologists can be investigation of the influence of current
management practices on physiological responses at the tree
and stand level, seeking to understand how stress may be altered and managed in economically justifiable ways.

The frequent complication of stress research by interactive effects of stress factors will surface throughout
this session. Simple theoretical relationships between environment and plant response often become more complex when
several factors of the environment act in concert. In the
laboratory, interactions can be controlled to some extent,
but in field research, scientists are often reduced to monitoring every conceivable environmental factor for later
correlation or covariance analysis, or must remain, to some
degree, uncertain about the operative conditions of their

experiments. Interactive effects on plant growth of two environmental stresses are often not additive; rather the relationships is complex and not easily described by mathematical functions. Compounding this problem is the fact that physiological responses to individual environmental stresses will shift as the length of stress imposition lengthens (i.e., time itself is an interacting factor in many supposed single factor responses). Hence in the field, stress interaction effects may complicate experiments enormously. This presents a special problem for modelers, where the influence of a complex environment must be simulated over long periods.

Finally, the occurrence of genetic variation in response to stress is another common theme of the session. It appears that "hereditary potential" for stress resistance to many environmental factors does indeed exist. This field offers much long-term potential for improving growth in forest stands, particularly in situations where environmental manipulation is not economically feasible and where genetic improvement can increase the level of response to management practices.

In summary, Klebs provides the best conceptual framework for this session: growth as a predictable behavior, a function of physiological processes in the plant, which are subject to genetic manipulation and a myriad of environmental influences.

REFERENCES

1. Blackman, F. F. 1905. Optima and limiting factors. Annals of Botany (London) 19: 281-295.
2. Klebs, G. 1913. Uber das Verhaltniss der Aussenwelt zur Entwicklung der Pflanzen. Sitzungsber. Heidelb. Akad. Wiss., Abt. B 5: 1-47.
3. Klebs, G. 1914. Uber das Treiben der einheimischen Baume, speziell der Buche. Sitzungsber. Heidelb. Akad. Wiss., Abh. Math.-Naturwiss. Kl. 3, reviewed in Plant World 18: 19, 1915.
4. Kramer, P. J. 1983. Water relations of plants. Academic Press, New York. 489 p.

8

5. Kramer, P. J., and T. T. Kozlowski. 1979. Physiology of woody plants. Academic Press, New York. 811 p.
6. Levitt, J. 1980. Responses of plants to environmental stresses. Vol. 2. Water, radiation, salt, and other stresses. Academic Press, New York. 607 p.
7. Lundegardh, H. 1939. Environment and plant development. Edward Arnold, London. 330 p.
8. Parker, W. C., S. G. Pallardy, T. M. Hinckley, and R. O. Teskey. 1982. Seasonal changes in tissue-water relations of three woody species of the Quercus-Carya forest type. Ecology 63: 1259-1267.
9. Salisbury, F. B., and C. W. Ross. 1978. Plant physiology. 2nd edition. Wadsworth Publishing, Belmont, California. 422 p.

2. MOISTURE: EFFECTS OF WATER STRESS ON TREES

R. O. TESKEY AND T. M. HINCKLEY
Assistant Professor, School of Forest Resources, University of Georgia, Athens, Georgia 30602 and Professor, College of Forest Resources, University of Washington, Seattle, Washington 98195.

ABSTRACT
Water stress is an important factor limiting the growth and productivity of forests. It can decrease growth directly through its effect on turgor, or indirectly by limiting carbon gain. Water availability can also alter the allocation of carbohydrates between the root and shoot. Trees resist excessive rates of water loss through stomatal regulation, a process which may be mediated by the level of growth regulators transported from the roots during periods of soil water deficits. An indirect effect of water stress on photosynthesis, due to a reduction of gaseous diffusion caused by stomatal closure, often has been noted. However, stress effects in the mesophyll may be more important limitations to photosynthesis than that caused by decreased diffusion of CO_2 into the leaf. Responses to water stress, including osmotic adjustment, are also discussed.

2.1 INTRODUCTION

A lack of moisture has a dramatic impact on cell, tree, and stand processes. Moisture stress affects growth and productivity of forests, (discussed by S. Pallardy, Chapter 1), and occurs when water content decreases to a level that affects physiological processes. At the cellular level, moisture stress can inhibit enzyme activity, affect membrane conformation, and influence all other physiological processes. At the level of the tree, it can decrease diameter and height growth, reduce the ability of the tree to resist other stresses, affect carbohydrate partitioning, and

influence the timing and rate of other physiological
processes, e.g., flower and fruit production. At the stand
level, it can cause a decrease in leaf area, an increase in
mortality, or the replacement of one species by another.
Such stressful conditions can occur during the year (e.g.,
summer drought, winter desiccation), during the day (e.g.,
high evaporative demand), and under both soil moisture
deficiencies and excesses.

The importance of moisture stress in regulating the
growth and development of forests has been apparent for many
years. The early work of Waring and Cleary (62) demonstrated
the ecological role of water in defining the distribution of
major community types in the Siskiyou Mountains of
southwestern Oregon (Figure 1). Similarly, Zahner (64)
portrayed the impact of irrigation on height, diameter, and
volume growth of Pinus taeda trees (Figure 2). These figures

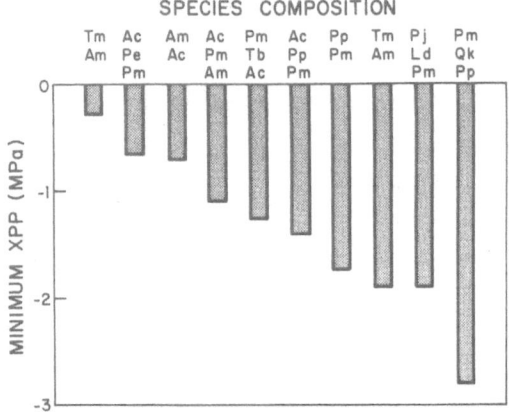

FIGURE 1. Minimum xylem pressure potentials (XPP) of
Pseudotsuga menziesii or Abies magnifica var. shastensis
saplings in various stands in the Siskiyou Mountains during a
drought. Dominant vegetation of each stand is listed in
order of importance. Abbreviations: Ac, Abies concolor; Am,
A. magnifica var. shastensis; Ld, Libocedrus decurrens; Pc,
Picea engelmannii; Pj, Pinus jeffreyi; Pp, P. ponderosa; Pm,
Pseudotsuga menziesii; Qk, Quercus kelloggii; Tb, Taxus
brevifolia; T. Tsuga mertensiana. Drawn from data presented
in (62)..

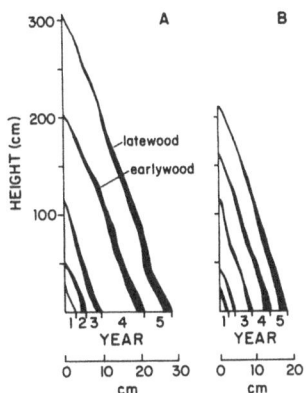

FIGURE 2. Stem growth of irrigated (A) and droughted (B)
Pinus taeda saplings. Treatments were applied in years 4 and
5. Redrawn from (64).

illustrate the relationships between water availability,
growth, and species distribution. In short, water deficits,
regardless of cause, result in growth reductions, impact
forest productivity, and affect species survival and
distribution.

In this chapter, we will review water stress and its
mode of action in affecting processes within the tree. We
will describe the extent to which moisture stress reduces
stand productivity and finally, we will identify key future
research needs.

2.2 DEFINING WATER STRESS

Any decrease in water content may be termed a water
deficit. When a decrease in water content, or increase in
water deficit, reaches a level which negatively affects a
physiological process, the tree is experiencing water stress.
In Levitt's (30) terminology, the decrease in water content
is a strain and the resultant impact on a tree is a stress.
A tree's ability to resist the stress is then defined as
moisture stress resistance. Unfortunately, these terms have
been used interchangeably and the distinctions between water

content, water deficit, and water stress are blurred. But
·even when used in a defined, quantitative sense, the
relationships among them are complicated by differences among
species, tree part, age, and preconditioning.

The description of water status in energy terminology,
e.g., water potential, has become universally accepted. This
terminology provides the means for defining water status
either within a tree or within the soil-tree-atmosphere
continuum and allows for quantitative comparisons. Water
potential has been defined (46) by the following equation:

$$\Psi = \Psi_p + \Psi_\pi + \Psi_\tau + \Psi_g \tag{1}$$

where Ψ is water potential and Ψ_p, Ψ_π, Ψ_τ, and Ψg are the
pressure, osmotic, matric, and gravitational potentials,
respectively. Pressure potential describes the pressure
against the cell wall and is determined by water volume and
the elasticity of the cell wall. Osmotic potential is a
measure of osmotically active solutes in the cell. The SI
unit of Ψ is the Pascal (1 MegaPascal = 10 bars). Frequently
the Ψ_τ and Ψ_g terms are omitted from measurement or
discussion since they are assumed to be small, although Ψ_g
can be important in tall trees (43, 54). The reduced
equation:

$$\Psi = \Psi_p + \Psi_\pi \tag{2}$$

is useful for describing the state of water in cells and
tissues. As shown in Figure 3, this equation can be
graphically presented (20, 24, 37, 52). This diagram
illustrates how the variables of equation 2 change during
dehydration or rehydration of tree tissues. Typically, data
used to develop this type of diagram are derived using the
pressure-volume curve technique (57, 58). As illustrated in
Figure 3, when water content decreases (i.e., through
dehydration or internal redistribution), 1) osmotic potential
decreases as cell solutes are concentrated, 2) pressure

13

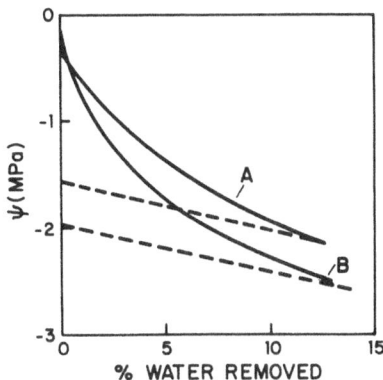

Figure 3. Pressure-volume curves for shoots of drought
stressed (B) and unstressed (A) Tsuga heterophylla seedlings.
Water potential is shown by the solid line, osmotic potential
by the dashed line. The vertical difference between the two
lines is the turgor potential (Ψ_T). Units of Ψ_T are
positive. The shift in osmotic potential between treatments
A and B represents osmotic adjustment. Redrawn from (24).

potential (or turgor pressure) drops as a function of cell
wall elasticity and the relative water content at which zero
turgor occurs, and 3) water potential decreases. Initially,
most of the decrease in water potential is caused by a
decrease in the pressure potential but after the point of
zero turgor, it is caused by changes in osmotic potential.

2.3 WATER DEFICIT FORMATION

Water deficits occur, or water potential decreases,
whenever there is a decrease in water content. Water
deficits may form because of changes in 1) atmospheric
evaporative demand, 2) soil moisture, 3) absorption of water
by the roots, 4) redistribution of water within tissues and
5) extracellular or extra-organ ice formation (41).
A useful equation for synthesizing our discussion of water
deficit formation was proposed by Richter (36) based upon the
work of Huber (21):

$$\Psi_x = \Psi_{soil} + \Psi_g + \sum_{i=soil}^{x} f_i \cdot r_i \qquad (3)$$

where Ψ_x is the water potential at some point in the tree and

$$\sum_{i=soil}^{x} f_i \cdot r_i$$

is the sum of the product of the partial fluxes (f_i) and resistances (r_i) between the soil and point x in the tree. Most authors assume $\Psi_g = 0$ in equation 3. As shown in equation 4, the components can be altered to simplify equation 3 further (8, 13, 18, 26).

$$\Psi_{leaf} = \Psi_{soil} + f_{transpiration} \cdot r_{soil \ to \ leaf} \qquad (4)$$

Although this equation elegantly illustrates the interaction between a number of variables (such as transpiration and soil water potential) and their impact on foliar water potential, it contains a number of weaknesses. It is frequently used with the assumption that a rather simplified flow path is sufficient to describe water flow through a tree and that there is no storage of water within tissues. The hysteresis loops shown in Figure 4 are examples of the effects of sapwood water storage or capacitance on the behavior of water potential during conditions of dehydration and subsequent rehydration.

As originally noted by Richter (36) and shown in Figure 5, the path of flow between the roots and foliage is not a simple single path, nor are the resistances ($\cong 1$/leaf specific conductivity) constant along the path. At each junction (i.e., branch to stem, branch to branch or leaf to branch) there is a reduction in diameter of the conduction elements (22, 28). This causes an increase in resistance which has been documented for Abies balsamea (10), Acer saccharum (65), Betula papyrifera (65) and Picea sitchensis (16). Zimmermann (66) hypothesized that because of the greater resistance to flow associated with junctions, a priority system of water,

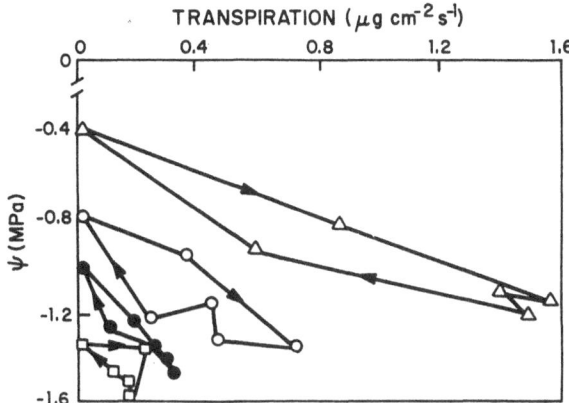

FIGURE 4. Daily course of water potential (Ψ) and transpiration in <u>Pinus contorta</u> trees for July 6 (Δ), July 26 (o), August 8 ($\bullet$) and August 30, 1978 ($\square$). Arrows indicate the sequence of samples. Diurnal hysteresis indicates that stored water from the sapwood is the initial source of transpired water. Data from (38).

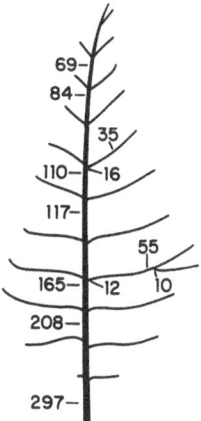

FIGURE 5. Leaf specific conductivities (hydraulic conductivity per gram dry weight of supplied leaves) of a 19-year-old <u>Tsuga canadensis</u>. Note the low conductivities at branch insertions on the main stem. From (10).

and hence nutrient distribution, was physically established. In addition, the resistances to flow through the xylem change both diurnally and seasonally. Measurement of stem diameter fluctuations (18, 29) and stem water content (63) suggest that these changes reflect changes in xylem water content.

Vascular cavitations, an indirect measure of decreases in
xylem water content, have also been shown to occur on both a
diurnal and seasonal basis (56, 59).

The resistance to water flow across the root is a
function of the viscosity of water, the state of the cell
membrane, and the anatomy and degree of suberization of the
tissue. Water flow to the root through the soil is by mass
flow following a water potential gradient from the soil to
the root. The rate of flow is dependent upon the magnitude
of the water potential gradient, soil hydraulic conductance,
temperature, and the degree of contact between the root and
soil. Figure 6 illustrates the effect of decreasing soil
temperature on the resistance to water flow in a variety of
woody species (25, 40, 52).

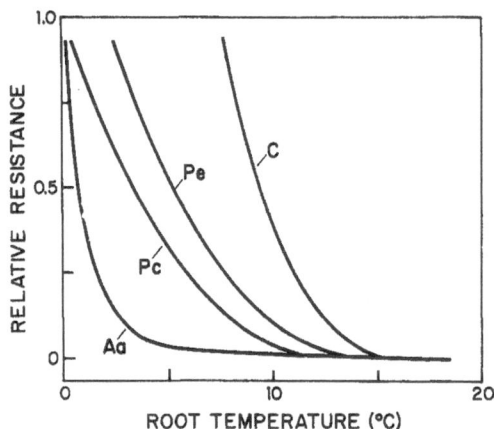

FIGURE 6. Relative resistance to water flow in Abies
amabilis (Aa), Pinus contorta (Pc), Picea engelmanii (Pe),
and Citrus sp. (C) at root temperatures less than 20°C.
Drawn from data in (25, 40, 52).

Changes in root properties, through decreasing soil
moisture or soil temperature, have both a direct effect
(described above) and an indirect effect on water flow
through plants. For example, Blackman and Davies (1) noted
that decreased root growth due to dry soil reduced cytokinin

production and resulted in stomatal closure. Low soil oxygen appears to "mimic" the effects of low soil temperature on root water absorption, root growth and cytokinin/abscisic acid production (49, 61).

In summary, water deficits are formed when transpiration exceeds the water supply available to the leaf. The water supply pathway can be considered as a complex series of resistances which, in woody plants, also contains a stored water component. The resistances in this pathway are not constant, but instead vary with soil environmental conditions, root distribution, xylem anatomy, shoot architecture and stomatal aperture. The evaporative demand is also important in the development of water deficits since, in the absence of stomatal closure, it controls the rate of transpiration.

2.4 IMPACT OF WATER STRESS ON PROCESSES

Water stress can affect processes within a tree either directly or indirectly. Water stress can directly affect growth by changing turgor related cell expansion as stated in the following equation:

$$\frac{dv}{dt} = W_{ex} \; (\Psi_p - \Psi_p^\gamma) \tag{5}$$

where W_{ex} is the cell wall extensibility, Ψ_p is the turgor pressure and Ψ_p^γ is the threshold turgor pressure for cell expansion. The role of turgor in growth was classically shown by Meyer and Boyer (32). Small decreases in water potential (hence, turgor – see Figure 3) resulted in dramatic decreases in leaf growth of Glycine max. At a particular leaf water potential, well above the water potential necessary for stomatal closure and subsequent decreases in net photosynthesis, leaf growth ceased. This point defined the Ψ_p^γ given in equation 5. Preliminary data for woody plants (60) suggested turgor related decreases in growth; however, growth reductions were found to also be related to

drought induced reductions in W_{ex} (Figure 7). In their
study, drought reduced cell wall extensibility, so that
leaves of drought stressed trees had stiffer cell walls.

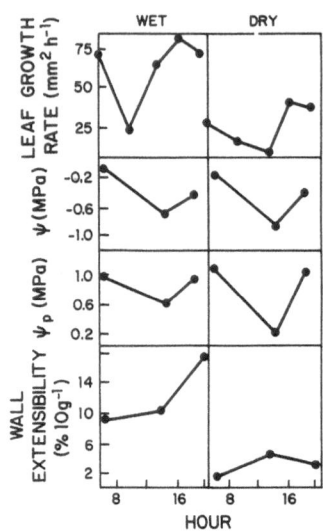

FIGURE 7. The relationship between time of day and cell wall
extensibility (% change for a 10 g strain), turgor pressure
(Ψ_p), leaf water potential (Ψ), and leaf growth for leaves
for hybrid 11 (<u>Populus</u> <u>trichocarpa</u> x <u>P</u>. <u>deltoides</u>) growing in
an irrigated and non-irrigated site at Puyullap, WA. Data
from August 24, 1984 (60).

In addition to the direct impact of water stress on cell
elongation or expansion, water stress can also affect other
physiological processes either directly or indirectly. Since
the work of Brix (3), a strong correlation has been noted
between changes in transpiration and photosynthesis (Figure
8). It has been often assumed that such a correlation
results from stomatal limitations to diffusion of CO_2 into
the leaf mesophyll. However, when stomatal and nonstomatal
limitations to photosynthesis are compared, the stomatal
limitation is often quite small (23). For <u>Pinus</u> <u>taeda</u>
seedlings, where the decrease in net photosynthesis mimicked

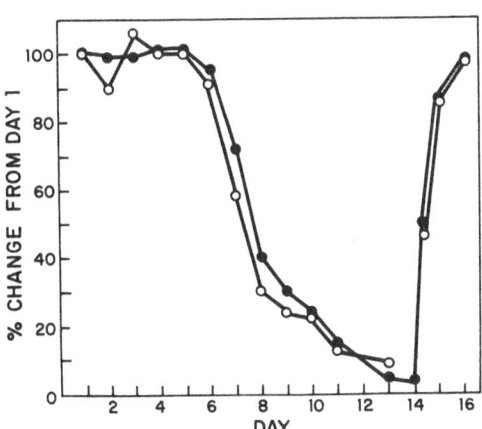

FIGURE 8. Effect of a drying cycle on the net photosynthesis (•) and transpiration (o) of <u>Pinus</u> <u>taeda</u> seedlings. Plants were rewatered on day 14. Redrawn from (3).

the decline in leaf conductance (Figure 9), the estimated gas phase limitation under well watered conditions was only 24% (i.e., mesophyll limitations contributed 76%) (53). As stress increased, the estimated gas phase limitation increased to 39%, and then declined to only 12% at the point of stomatal closure (-2.0 MPa). This suggested that direct inhibition of net photosynthesis had occurred at low leaf water potentials. Nonstomatal inhibition can also be clearly seen in trees with poor stomatal regulation (Figure 10). In these plants, net photosynthesis declines with leaf water potential, yet no correlation exists between leaf conductance and net photosynthesis (Scarascia, unpublished data). This response may be due to the effect of decreasing turgor in the mesophyll (affecting spatial relationships between parts of cells) and reduction in activity of membrane bound enzymes. If stomatal limitations are relatively low, then it is likely that water stress directly affects mesophyll processes to a much greater degree than originally thought. More data are

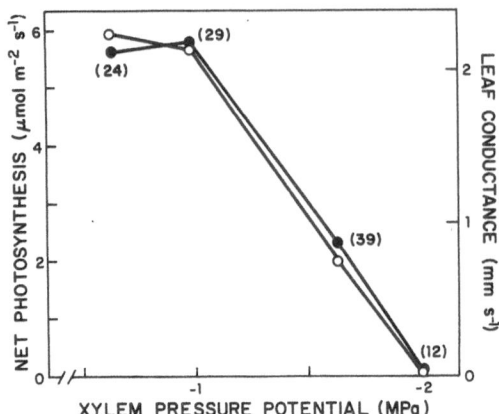

FIGURE 9. Net photosynthesis (●) and leaf conductance to water vapor (o) in <u>Pinus</u> <u>taeda</u> seedlings exposed to a drying cycle. Numbers in parenthesis indicate the percent gas phase limitation to net photosynthesis (primarily stomatal) during an imposed drought. Data from (53).

needed on this relationship before a conclusion can be reached.

Regardless of the nature of the primary limit to photosynthesis, photosynthesis and stomatal conductance tend to simultaneously increase or decrease (Fig. 8, 9). Factors related to moisture stress which affect stomatal aperture include 1) leaf water potential, 2) predawn leaf water potential, 3) absolute humidity difference between leaf and air and 4) the level of plant growth substances such as abscisic acid and cytokinin. There is often a consistent pattern in the response of the stomata of trees to declining leaf water potentials. Usually there is an initial range of water potentials which cause little or no reduction in stomatal conductance or net photosynthesis (Fig. 8, 9). If leaf water potential continues to decline, a point is reached where a rapid decrease in stomatal conductance occurs. The water potential at which this happens is termed the water

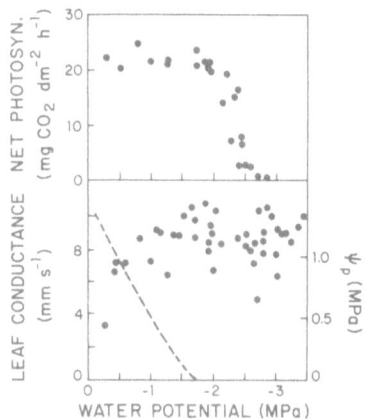

FIGURE 10. Effect of decreasing water potential on net photosynthesis, leaf conductance to water vapor and pressure potential (Ψ_p, dashed line) in <u>Populus</u> <u>trichocarpa</u> (G. Scarascia, unpublished data).

potential threshold. A range of threshold leaf water potentials for tree species is shown in Table 1.

Predawn leaf water potential is also correlated with stomatal conductance. For example, maximum daily conductance or average daily conductance can be related to it (Fig. 11) (19, 39). This correlation exists because predawn leaf water potential is an excellent index of soil water availability, and even root growth activity in some species (50). It not only reflects soil water potential, but also integrates rooting area and localized rhizosphere resistances, which are all key components in equation 4.

Evaporative demand affects stomatal conductance as well. Large absolute humidity deficits may cause stomatal closure directly by reducing turgor in the guard cells (55). Large humidity deficits can also reduce leaf water potential since under these conditions the rate of transpiration is much greater than the rate of water movement into the leaf. If the leaf water potential threshold is surpassed, then closure will result. Many tree species react to large absolute humidity deficits by partially closing their stomata, which

TABLE 1. Threshold leaf water potentials for the initiation
of stomatal closure for selected species.

Species	Leaf water potential	
Abies amabilis	-1.5 MPa	(52)
Abies grandis	-1.2	(38)
Acer saccharum	-1.7	(5)
Picea engelmannii	-1.0	(25)
Pinus ponderosa	-1.8	(38)
Pinus taeda	-0.9	(53)
Pseudotsuga menziesii	-1.6	(38)
Quercus alba	-2.3	(5)
Quercus rubra	-1.9	(5)
Quercus velutina	-2.5	(5)

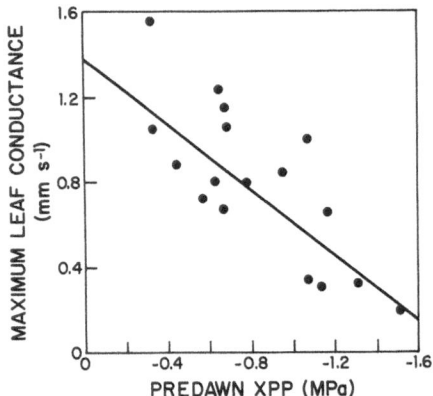

FIGURE 11. Relationship between maximum leaf conductance and
predawn xylem pressure potential (XPP) for Pseudotsuga
menziesii trees. Each point represents at least nine leaf
conductance measurements. Redrawn from (39).

conserves water (19). Although photosynthesis is reduced,
closure may be necessary to prevent damage to the leaf cells

·from desiccation. The relation between soil water
availability, evaporative demand, and stomatal conductance is
demonstrated in Figure 12.

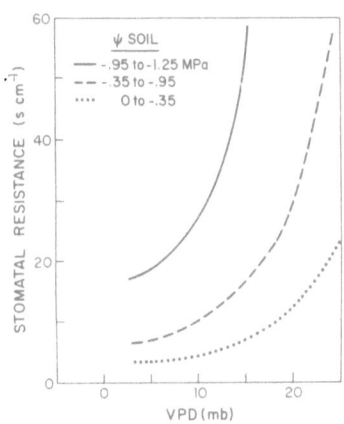

FIGURE 12. Relationship between stomatal resistance
(1/conductance) and vapor pressure deficit (VPD) at three
soil water potential levels (Ψ_{soil}) for <u>Psuedostuga menziesii</u>
trees. The increasing steepness in slope under drier soil
conditions implies increasing resistance to flow from soil to
leaf. Redrawn from (48).

2.5 RESPONSES TO DROUGHT

     Exposure to water stress can also produce reactions
within the tree which may help it either resist future
drought episodes, or reduce the impact of the present
episode. These include (from the short to the long term)
1) physical or biochemical "sensing" of rhizosphere stresses,
2) osmotic adjustment, 3) foliage or branch abscission, and
4) increased fine root production. Recent evidence indicates
that the shoot can "sense" stresses in the rhizosphere,
including water deficits, by means other than changes in leaf
water potential. The communication link between root and
shoot appears to be both short-term physical (51) and
mid-term biochemical stimuli (1, 2). Blake and Ferrell (2)
found that abscisic acid concentrations in the foliage of
<u>Pseudotsuga menziesii</u> greatly increased when soil water

potentials decreased. Coutts (6) and Teskey et al. (51)
reported that stresses on the root systems of conifer
seedlings caused changes in stomatal conductance which were
not associated with changes in bulk leaf water potential.
Recently, Blackman and Davies (1) have suggested that a
reduction in cytokinin produced by the roots may be the
factor responsible for this stomatal conductance response.
In their experiments using Zea mays, drying the soil around
part of the root system induced partial stomatal closure,
without changing leaf water potential. This closure was
reversible when the leaves were incubated with kinetin or
zeatin.

A longer-term response to water stress is an adjustment
of the root-shoot ratio. When exposed to drought the
allocation of carbohydrates to root growth may increase,
providing more root absorptive area per unit area of foliage
and increasing the volume of soil explored. However, since
plant size is a confounding factor in many of these studies,
the data should be interpreted cautiously. Another response
to drought is a change in the osmotic potential of the cells,
termed osmotic adjustment. The advantages of osmotic
adjustment are clear: lowering the osmotic potential will
1) increase the water potential gradient from soil to leaf,
allowing the plant to take up more water and to extract water
from soil held at lower water potentials, and 2) maintain a
given turgor pressure at lower water potentials. An example
of osmotic adjustment is shown in Figure 3 for Tsuga
heterophylla seedlings.

The degree of osmotic adjustment in response to drought
appears to vary with species, genotype, and degree of stress.
Parker and Pallardy (33) found that after a single drought
cycle, Juglans nigra seedlings from an Ontario genotype
shifted leaf osmotic potentials by as much as -0.8 MPa, but a
New York source had no detectable adjustment. Seiler (45)
reported an osmotic adjustment of 0.4 MPa in Alnus glutinosa
seedlings which were repeatedly droughted over a 12 week
period. Buxton et al. (4) found that osmotic potentials

decreased approximately -0.3 MPa when seedlings of Picea
glauca, P. mariana, and Pinus banksiana were exposed to a
mild water stress caused by -0.4 MPa solution of polyethylene
glycol. However, more negative osmotic solutions (-0.8 and
-1.6 MPa) did not cause a further decrease in leaf osmotic
potential. The osmotic potentials of these seedlings also
returned to prestress levels within 72 hours after relief of
the stress.

Seasonal changes in leaf osmotic potential have been
noted in a number of species (Abies amabilis (52); Abies
homolepis, Betula ermani, Quercus crispula (31); Carya
tomentosa, Quercus alba, Q. rubra (34); Picea abies (15); and
Pseudotsuga menziesii (35). While this is often considered
osmotic adjustment, these changes may only represent leaf
maturation (i.e., accumulation processes rather than active
adjustment), and other environmental stresses, such as low
temperatures, are likely to be involved. The influence of
temperature is particularly important in conifers. Ritchie
and Shula (35), Gross et al. (15), Teskey et al. (52), and
Doi et al. (7) found more negative osmotic potentials in
conifer foliage in the fall and winter than in the summer.
Hennessey and Dougherty (17) recently reported that the
osmotic potentials of Pinus taeda seedlings grown in a
nursery decreased 0.4 MPa when compared with well watered
controls after being exposed to drought and winter
temperatures. During the summer and fall there were no
significant differences in osmotic potentials between
treatments. These results suggest that a number of
environmental factors may induce shifts in osmotic potential.

2.6 WATER STRESS AND STAND PRODUCTIVITY

Water availability affects stand productivity through
its effect on 1) leaf area, 2) photosynthesis and 3)
carbohydrate allocation patterns. The effect of water
availability on leaf area is probably the most important
means by which water stress influences productivity. Grier
and Running (14) reported that the maximum leaf area of

mature forests in the Pacific Northwest was a linear function of site water balance (Figure 13). Differences in the leaf area index were dramatic: from approximately 38 m^2 m^{-2} for Picea sitchensis growing in a moist coastal environment to 6 m^2 m^{-2} for Juniperus occidentalis on a site with very low precipitation.

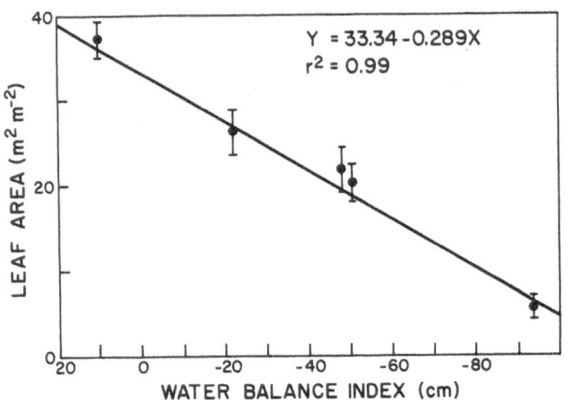

FIGURE 13. Relationship between total leaf area of mature forest stands along a transect of decreasing site water balance. Dominant species in each stand (from left to right) are: Tsuga heterophylla and Picea sitchensis; Pseudotsuga menziesii and Tsuga heterophylla; Pseudotsuga menziesii and Quercus garryana; Pseudotsuga menziesii, Abies grandis and Pinus ponderosa; and Juniperus occidentalis (14).

Using a similar transect, Gholz (12) recently found a strong correlation between overstory net primary production and leaf area index. Similar correlations have been developed for plantations of Pinus densiflora (Figure 14) (42). These relationships indicate the importance of photosynthetic surface area in determining productivity, and the sensitivity of stand leaf area to the environment. Emmingham (9) noted a good correlation between stand basal area and an index of maximum moisture stress (minimum predawn xylem pressure potential) for a number of western conifers. Using another index of water availability, Fralish et al. (11) observed a linear relationship between basal area,

species distribution, and soil water availability for
deciduous hardwoods. Although these relationships
demonstrate the influence of water on forest productivity, it
should be remembered that adequate water availability is
necessary for, but will not guarantee, high productivity,
because other environmental factors also affect carbon gain -
especially nutrient status and temperature.

Although it is well known that water stress reduces net
photosynthesis, the correlations between net photosynthesis
and productivity are poor. This is likely to always be the
situation until point measurements of photosynthesis can be

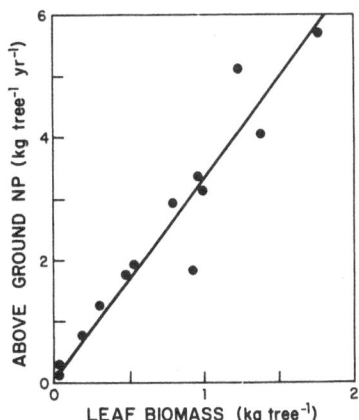

FIGURE 14. Above ground net production (NP) aⁱₐ leaf biomass
of <u>Pinus</u> <u>densiflora</u> trees in a 15-year-old forest (42).

sampled at a sufficient frequency to provide representative
estimates of canopy photosynthesis and canopy carbon gain.

As mentioned earlier, water can also alter aboveground
productivity by causing changes in root-shoot allocation
patterns. It has been hypothesized that a plant responds to
soil moisture deficits by increasing the size of the root
system, thereby increasing surface area and absorptive
capacity (44). While as yet unproven, there is some

experimental data to substantiate this hypothesis for annual plants, and there is some circumstantial evidence to support it for tree species. For example, Teskey and Hinckley (50) noted that root growth of Quercus alba trees increased as soil water potential decreased from -0.05 to -0.8 MPa. It is likely that changes in allocation patterns are general for all limitations which exist in the soil. Keyes and Grier (27) showed that two Pseudotsuga menziesii stands of the same age which were growing on what appeared to be sites of "high" and "low" productivity actually had very similar annual net primary productivities. However, on the "low productivity site" a much greater proportion of the available carbohydrates was allocated for fine root growth to the detriment of shoot growth. It is probable that this response was due to both water and nutrient limitations.

In conclusion, adequate water availability is essential for high productivity. It affects the leaf area of the stand and the net photosynthetic rate of the foliage. An adequate water supply may also result in a higher proportion of aboveground growth by reducing root system development.

ACKNOWLEDGEMENTS: We thank Dr. B. Smit-Spinks for her critical review of this chapter and Dr. G. Scarascia for graciously allowing us the use of unpublished data. This review was supported in part by a McIntire-Stennis Grant to ROT.

REFERENCES

1. Blackman, P. G. and W. J. Davies. 1985. Root to shoot communication in maize plants of the effects of soil drying. J. Exp. Bot. 36:39-48.
2. Blake, J. and W. K. Ferrell. 1977. The association between soil and xylem water potential, leaf resistance and abscisic acid content in droughted seedlings of Douglas-fir (Pseudotsuga menziesii). Physiol. Plant. 39:106-109.
3. Brix, H. 1962. The effect of water stress on the rates of photosynthesis and respiration in tomato plants and loblolly pine seedlings. Physiol. Plant. 15:10-20.

4. Buxton, G. F., D. R. Cyr, E. B. Dumbroff and D. P. Webb. 1985. Physiological responses of three northern conifers to rapid and slow induction of moisture stress. Can. J. Bot. 63:1171-1176.
5. Chambers, J. L., T. M. Hinckley, G. S. Cox, C. L. Metcalf and R. G. Aslin. 1985. Boundary-line analysis and models of leaf conductance for four oak-hickory forest species. For. Sci. 31:437-450.
6. Coutts, M. P. 1980. Control of water loss by actively growing Sitka spruce seedlings after transplanting. J. Exp. Bot. 31:1587-1597.
7. Doi, K., Y. Morikawa and T. M. Hinckley. 1986. Seasonal trends of several water relations parameters in Cryptomeria japonica seedlings. Can. J. For. Res. (in press).
8. Elfving, D. C., M. R. Kaufmann and A. E. Hall. 1972. Interpreting leaf water potential measurements with a model of the SPAC. Physiol. Plant. 27:771-780.
9. Emmingham, W. H. 1982. Ecological indexes as a means of evaluating climate, species distribution and primary production. In: (R. L. Edmonds, ed.) Analysis of Coniferous Forest Ecosystems in the Western United States. US/IBP Synthesis Series 14. Hutchinson Ross Publ. Co., Stroudsburg, Pennsylvania. 45-67.
10. Ewers, F. W. and M. H. Zimmermann. 1984. The hydraulic architecture of eastern hemlock (Tsuga canadensis). Can. J. Bot. 62:940-946.
11. Fralish, J. S., S. M. Jones, R. K. O'Dell and J. L. Chambers. 1978. The effect of soil moisture on site productivity and forest composition in the Shawnee Hills of southern Illinois. In: (W. E. Balmer, ed.) Proceedings Soil Site Productivity Symposium. USDA For. Ser., Washington, D.C. 263-285.
12. Gholz, H. L. 1982. Environmental limits on aboveground net primary production, leaf area, and biomass in vegetation zones of the Pacific Northwest. Ecology 63:469-481.
13. Ginter-Whitehouse, D. L., T. M. Hinckley and S. G. Pallardy. 1983. Spatial and temporal aspects of water relations of three tree species with different vascular anatomy. For. Sci. 29:317-329.
14. Grier, C. C. and S. W. Running. 1977. Leaf area of mature northwestern coniferous forests: Relation to site water balance. Ecology 58:893-899.
15. Gross, V. K., T. Pham-Nguyen, H. Unger. 1980. Tägliche und Saisonale änderungen des wasserpotentials und siener Komponenten in den Kronen von Fichten underschiedlichen Alters. Allg. Forst. Jagdztg. 151:69-80.
16. Hellkvist, J., G. P. Richards and P. G. Jarvis. 1974. Vertical gradients of water potential and tissue water relations in Sitka spruce trees measured with the pressure chamber. J. Appl. Ecol. 11:637-668.

17. Hennessey, T. C. and P. M. Dougherty. 1984.
Characterization of the internal water relations of
loblolly pine seedlings in response to nursery cultural
treatments: Implications for reforestation success.
In: (M. L. Duryea and G. N. Brown, eds.) Seedling
Physiology and Reforestation Success. 225-246.

18. Hinckley, T. M. and D. M. Bruckerhoff. 1975. The
effects of drought on water relations and stem shrinkage
of Quercus alba. Can. J. Bot. 53:62-72.

19. Hinckley, T. M., J. P. Lassoie and S. W. Running. 1978.
Temporal and spatial variations in the water status of
forest trees. For. Sci. Monogr. 20. 72 pp.

20. Höfler, K. 1920. Ein Schema für osmotische Leitung der
Pflanzenzelle. Berl dtsch bot Ges 35:707-726.

21. Huber, B. 1928. Weitere quantitative Untersuchungen
über das Wasserleitungssystem der Pflanzen. Jahrb. Wiss.
Bot. 67:877-959.

22. Isebrands, J. G. and P. R. Larson. 1977. Vascular
anatomy of the nodal region in eastern cottonwood. Am.
J. Bot. 64:1066-1077.

23. Jones, H. G. 1985. Partitioning stomatal and
non-stomatal limitations to photosynthesis. Plant, Cell
and Environ. 8:95-104.

24. Kandiko, R. A., R. Timmins and J. Worrall. 1980.
Pressure-volume curves of shoots and roots of normal and
drought conditioned western hemlock seedlings. Can. J.
For. Res. 10:10-16.

25. Kaufmann, M. R. 1975. Leaf water stress in Engelmann
spruce: Influence of the root and shoot environments.
Plant Physiol. 56:841-844.

26. Kaufmann, M. R. 1979. Stomatal control and the
development of water deficit in Engelmann spruce
seedlings during drought. Can. J. For. Res. 9:297-304.

27. Keyes, M. R. and C. C. Grier. 1981. Above- and
below-ground net production in 40-year-old Douglas-fir
stands on low and high productivity sites. Can. J. For.
Res. 11:599-605.

28. Larson, P. R. and J. G. Isebrands. 1978. Functional
significance of the nodal constricted zone in Populus
deltoides. Can. J. Bot. 56:801-804.

29. Lassoie, J. P. 1973. Diurnal dimensional fluctuations
in a Douglas-fir stem in response to tree water status.
For. Sci. 19:251-255.

30. Levitt, J. 1980. Responses of Plants to Environmental
Stresses. Vol. II. Water, Radiation, Salt and other
Stresses. Academic Press, New York. 607 pp.

31. Maruyama, Y. and Y. Morikawa. 1984. Seasonal changes
of several water relations parameters in Quercus
crispula, Betula ermani and Abies homolepis. J. Jap.
For. Soc. 66:499-505.

32. Meyer, R. F. and J. S. Boyer. 1972. Sensitivity of
cell division and cell elongation to low water
potentials in soybean hypocotyls. Planta 108:77-78.

33. Parker, W. C. and S. G. Pallardy. 1985. Stem vascular anatomy and leaf area in seedlings of six black walnut (Juglans nigra) families. Can. J. Bot. 63:1266-1270.

34. Parker, W. C., S. G. Pallardy, T. M. Hinckley and R. O. Teskey. 1982. Seasonal changes in tissue water relations of three woody species of the Quercus-Carya forest type. Ecology 63:1259-1267.

35. Ritchie, G. A. and R. G. Shula. 1984. Seasonal changes of tissue-water relations in shoots and root systems of Douglas-fir seedlings. For. Sci. 30:538-548.

36. Richter, H. 1973. Frictional potential losses and total water potential in plants: A re-evaluation. J. Exp. bot. 24:983-994.

37. Roberts, S. W. and K. R. Knoerr. 1977. Components of water potential estimated from xylem pressure measurements in five species. Oecologia 28:191-202.

38. Running, S. W. 1976. Environmental control of leaf water conductance in conifers. Can. J. For. Res. 6:104-112.

39. Running, S. W. 1980. Environmental and physiological control of water flux through Pinus contorta. Can. J. For. Res. 10:82-91.

40. Running, S. W. and C. P. P. Reid. 1980. Soil temperature influences on root resistance of Pinus contorta seedlings. Plant Physiol. 65:635-640.

41. Sakai, A. 1983. Comparative study on freezing resistance of conifers with special reference to cold adaptation and its evolutive aspects. Can. J. Bot. 61:2323-2332.

42. Satoo, T. 1968. Primary production relations in woodlands of Pinus densiflora. In: (H. E. Young, ed.) Symposium on Primary Productivity and Mineral Cycling in Natural Ecosystems. Univ. of Maine Press, Orono, Maine. 52-80.

43. Scholander, P. F., H. T. Hammel, E. D. Bradstreet and E. A. Hemmingsen. 1965. Sap pressure in vascular plants. Science 143:339-346.

44. Schulze, E. D., K. Schilling and S. Nagarajah. 1983. Carbohydrate partitioning in relation to whole plant production and water use of Vigna unguiculata (L.) Walp. Oecologia 58:169-177.

45. Seiler, J. R. 1985. Morphological and physiological changes in black alder induced by water stress. Plant. Cell and Environ. 8:219-222.

46. Slatyer, R. O. 1967. Plant-Water Relationships. Academic Press, New York.

47. Tan, C. S., T. A. Black and J. U. Nnyamah. 1977. Characteristics of stomatal diffusion resistance in a Douglas fir forest exposed to soil water deficits. Can. J. For. Res. 7:595-604.

48. Tan, C. S., T. A. Black and J. U. Nnyamah. 1978. A simple diffusion model of transpiration applied to a thinned Douglas-fir stand. Ecology 59:1221-1229.

32

49. Termaat, A., J. B. Passioura and R. Munns. 1985. Shoot turgor does not limit shoot growth of NaCl-affected wheat and barley. Plant Physiol. 77:869-872.
50. Teskey, R. O. and T. M. Hinckley. 1981. Influence of temperature and water potential on root growth of white oak. Physiol. Plant. 52:363-369.
51. Teskey, R. O., T. M. Hinckley and C. C. Grier. 1983. Effect of interruption of flow path on stomatal conductance of Abies amabilis. J. Exp. bot. 34:1251-1259.
52. Teskey, R. O., C. C. Grier and T. M. Hinckley. 1984. Change in photosynthesis and water relations with age and season in Abies amabilis. Can. J. For. Res. 14:77-84.
53. Teskey, R. O., J. A. Fites, L. J. Samuelson and B. C. Bongarten. 1986. Environmental influences to net photosynthesis in Pinus taeda (L.) seedlings. Tree Physiol. (in press).
54. Tobiessen, P., P. W. Rundel and R. E. Stecker. 1971. Water potential gradient in a tall Sequoiadendron. Plant Physiol. 48:303-304.
55. Turner, N. C., E. D. Schulze and T. Gollan. 1985. The responses of stomata and leaf gas exchange to vapor pressure deficits and soil water content. II. In the mesophytic herbaceous species Helianthus annuus. Oecologia 65:348-355.
56. Tyree, M. T. and M. A. Dixon. 1983. Cavitation events in Thuja occidentalis L.? Plant Physiol. 72:1094-1099.
57. Tyree, M. T. and H. Richter. 1981. Alternative methods of analyzing water potential isotherms: some cautions and clarifications. J. Exp. bot. 32:643-653.
58. Tyree, M. T., N. S. Cheung, M. E. MacGregor and A. J. B. Talbot. 1978. The characteristics of seasonal and ontogentic changes in the tissue-water relations of Acer, Populus, Tsuga and Picea. Can. J. Bot. 56:635-647.
59. Tyree, M. T., M. E. D. Graham, K. E. Cooper and L. J. Bazos. 1983. The hydraulic architecture of Thuja occidentalis. Can. J. Bot. 61:2105-2111.
60. Van Volkenburgh, E., C. Ridge and T. M. Hinckley. 1985. Limits to poplar leaf growth. Plant Physiol. (Suppl.) 77:136.
61. Wadman-van Schravendijk, H. and O. M. van Andel. 1985. Interdependence of growth, water relations and abscisic acid level in Phaseolus vulgaris during waterlogging. Physiol. Plant. 63:215-220.
62. Waring, R. H. and B. D. Cleary. 1967. Plant moisture stress: evaluation by pressure bomb. Science 155:1248-1254.
63. Waring, R. H. and S. W. Running. 1978. Sapwood water storage: Its contribution to transpiration and effect upon water conductance through the stems of old growth Douglas fir. Plant, Cell and Environ. 1:131-140.

64. Zahner, R. 1962. Terminal growth and wood formation by juvenile loblolly pine under two soil moisture regimes. For. Sci. 8:345-352.
65. Zimmermann, M. H. 1978. Hydraulic architecture of some diffuse-porous hardwoods. Can. J. Bot. 56:2286-2295.
66. Zimmermann, M. H. 1983. Xylem Structure and the Ascent of Sap. Springer-Verlag, Berlin. 143 pp.

3. MOISTURE-STRESS MANAGEMENT: SILVICULTURE AND GENETICS

R. J. NEWTON[1], C. E. MEIER[1], J. P. VAN BUIJTENEN[2] AND
C. R. MCKINLEY[2]

[1]Associate Professor and Assistant Professor, Department of
Forest Science and the Texas Agricultural Experiment Station,
Texas A&M University, College Station, TX 77843.

[2]Professor and Assistant Professor, Department of Forest
Science and the Texas Forest Service, Texas A&M University,
College Station, TX 77843.

ABSTRACT

Silvicultural treatments and the development of site-
adapted genotypes can aid in the management of drought-
stress in forest species. Both pre- and post-planting
silvicultural treatments of the seedling and site have proven
effective in enhancing survival and growth under drought-
stress conditions. Silvicultural treatments modify the
stand and its environment and, consequently tree
physiological mechanisms in response to drought; this
modifies the water balance in forest stands. Evapotranspiration is directly related to stand leaf area.
A balance between water loss and productivity can be accomplished by use of silvicultural regimes which manipulate crop
leaf area and eliminate non-crop competing leaf area.

Intra-specific physiological differences in adaptation to
drought-stress have been identified. These differences
provide a basis for understanding the genetic expression of
drought resistance as well as serving as "markers" of
drought-resistant selections. Tissue culture evaluations
suggest that drought resistance is genetically expressed

Technical Article No. 21251, Texas Agricultural Experiment
Station, College Station, TX 77843

at the cellular level, and thus it may be a technique for
rapid selection for drought resistance. Field evaluations
for drought resistance are often unreliable, but greenhouse
and growth chamber tests have been successful. Site matching
with genetic strain and species can maximize productivity in
drought-prone areas.

3.1 INTRODUCTION

Growth of forest trees is greatly reduced by water
deficits (67). Zahner (67) has shown that 70-80% of the
variation in the width of annual rings can be attributed to
differences in precipitation patterns. In addition to
reduced growth, severe drought also reduces tree survival;
Williston (62) noted that in a southern watershed project
over a 16-year period, 57% of the first-year mortality in
pine plantations was due to drought.

Drought is defined as a sustained period of time without
significant rainfall or as prolonged dry weather (33). For a
period of dry weather to affect a forest community, the
rainfall deficit must lead to depletion of available soil
water reserves and ultimately to a tree water deficit (16).
The degree to which drought influences forest stand productivity depends on the magnitude and availability of soil
water reserves, the aridity (evaporative demand) of the
atmosphere, tree characteristics that influence water uptake,
the rate of transpiration, and the physiological responses of
trees to the drought-stress they are experiencing. Drought-
stress is a condition in which the cells are less than fully
turgid and the water potential is substantially less than
zero (30). Plants which can withstand site water deficits
and subsequent drought-stress are termed drought-resistant.

In order to understand how genetic improvements and
silvicultural treatments provide drought resistance and how
they may increase stand survival and productivity, it is
important to relate these topics to tree physiology.
Physiological mechanisms of drought resistance have been

characterized in forest species, and it has been shown that these mechanisms are modified by the genetic make-up of the tree (3, 4, 29, 37, 50, 53, 54) and silvicultural practices (20, 35).

3.2 MECHANISMS OF DROUGHT RESISTANCE

The term drought resistance is used to describe physiological and biochemical features that contribute positively to the ability of plants to survive, produce biomass, and reproduce in conditions of limited water availability (24). Two primary types of drought resistance have been identified in forest species (24, 32) (Table 1): (1) Drought tolerance at high tissue water potential: the ability of trees to endure periods of rainfall deficit while maintaining a high tissue water potential; (2) Drought tolerance at low tissue water potential: the ability of trees to endure rainfall deficits at low tissue water potential.

TABLE 1. Physiological mechanisms of drought resistance in forest species. Adapted from (24).

Drought Tolerance at High Tissue Water Potential (Avoidance)
1. Maintenance of water uptake
   a. Increased rooting
   b. Increased hydraulic conductance
2. Reduction of water loss
   a. Reduced epidermal conductance
   b. Reduced absorbed radiation
   c. Reduced evaporative surface
      i. reduced shoot size
      ii. reduced leaf size
Drought Tolerance at Low Tissue Water Potential (Tolerance)
1. Maintenance of turgor
   a. Solute accumulation
   b. Increased elasticity
2. Desiccation tolerance
   a. Protoplasmic resistance

Tolerance at high tissue water potential is often referred to as drought-avoidance, and tolerance at low tissue water potential is defined as drought-tolerance (32). Within these two types of drought resistance there are a number of mechanisms that enable trees to resist drought. These are shown in Table 1 which is modified from (24).

3.3 SILVICULTURAL MANIPULATION AND SITE DROUGHT-STRESS
3.3.1 The Biological Basis for Manipulation
     Silvicultural manipulation of site related water stress will be considered in relation to two phases of stand development: 1) stand establishment - the early survival and growth of trees prior to initial crown closure and 2) stand growth - the growth and development of the stand after crown closure. In the establishment phase, silvicultural emphasis is upon the individual tree. Practices are generally designed to maximize soil moisture reserves and the seedlings' exploitation of these reserves - with and without competing vegetation. After canopy closure, emphasis shifts away from the individual toward stand productivity. Survival of the individual is of lesser importance and, in fact, mortality of some trees is expected.
     The biological basis for the individual tree's response to drought stress and survival or establishment have been presented in Section 3.2 and summarized in Table 1. Relative to stand growth, perhaps the most important physiological mechanisms are those related to canopy water loss (epidermal conductance, absorbed radiation and evaporative surfaces). The following discussion will emphasize these mechanisms in regard to their biological bases for stand productivity and their linkage to within stand mortality and possible thinning regimes. In this discussion stand productivity will refer to the stand's aboveground annual wood increment.
     Jarvis (23) and Gordon et al. (15) have argued that production increment is directly related to the amount of photosynthetically active radiation intercepted by the canopy. Because leaf area is responsible for both loss of water by evapotranspiration and carbon fixation by photo-

synthesis, it forms a critical link between stand production and drought-stress. Jarvis (23) states that the largest effect of drought-stress on growth is through reduction of stand leaf area and thus interception of radiation. Decline in stand leaf area in response to increasing drought stress has proven to be a relatively consistent and sensitive relationship, despite a diversity in species and climatic regimes (12, 17, 43, 55, 56).

The optimuim leaf area for wood production can only be approximately defined. Waring estimated that maximum production efficiency (wood increment/unit leaf area/yr), an index of vigor, occurs at a leaf area index (LAI) of somewhat less than four (55). Jarvis (23) appears to agree with this estimate. However, the above LAI is not the LAI of maximum stand productivity (wood increment/ha/yr). Jarvis showed that at least theoretically a LAI of approximately 10 was needed to intercept 95% of the photosynthetically active radiation, and increasing stand productivity was assumed until this LAI was approached. Waring (55) working with actual stands, reported maximum stand production at a LAI of approximately six. Declines in production at higher LAI's were due to increasing stand respiration and mortality and thus lower net annual wood production (55). From these studies the optimium leaf area index for maximum annual wood production appears to be between 6 and 10. Optimum leaf area will undoubtedly vary with species and climate. Up to these LAI levels, assuming factors other than light are not limiting, increases in leaf area out-weigh decreases in per unit leaf efficiency.

In the above discussion of optimum LAI for maximum wood yield, drought-stress was not considered. As noted earlier, drought-stress may markedly decrease LAI. If maximum yield of commercial species is desired, then a silvicultural program should strive to bring the stand's LAI to as near the optimum LAI as possible and maximize the LAI of desired species. These goals are pursued through practices which modify the components of the site's water balance (23). For example, a stand's LAI is based on total leaf area; thus, if transpiring leaf area of the competing understory component is eliminated, crop tree leaf area may be expected to increase due to greater water availability. Alternatively, total leaf area may be increased by enhancing the site's rooting volume and soil moisture retention assuming there are no other more limiting factors.

The linkage between leaf area and silvicultural thinning regimes can be demonstrated in several ways. One which appears clear because of its link between leaf area or biomass and stand density is the "self-thinning rule". This rule was first introduced by Tadaki and Shidei (49, 59). The rule's characteristic equation is

$$B = CN^{-1/2}, \text{ or } W = CN^{-3/2},$$

where B = plant biomass per unit land area, N = density (stems per unit land area) of surviving plants, W = mean individual mass of survivors, and C = a constant. Since W = B/N, the two equations are equivalent (61). The basic concept is shown in Figure 1. The self-thinning rule describes plant mortality due to competition in crowded even-aged stands. A stand of a given density accumulates biomass until the self-thinning line is approached (Figure 1). Then the stand suffers mortality while continuing to accumulate biomass such that it moves along the self thinning line (61) until a maximum biomass characteristic of the species and site is attained. The above relationship, while more widely tested with biomass (B) and stand density (N), has also proven valid for leaf area and stand density (60). Thus, the interrelationship between leaf area dynamics and silvicultural practices, as suggested by Waring (55) and Jarvis (23) appears evident.

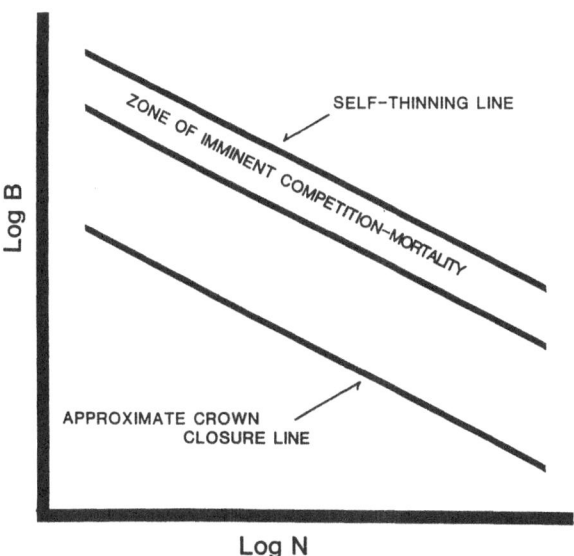

FIGURE 1. The self-thinning rule showing the relationship
between stand biomass and density of surviving
plants. Modified from (9, 61).

The linkage between silviculture and the self-thinning
rule has been demonstrated by Drew and Flewelling (8, 9)
(Figure 1). In addition to the primary self-thinning line,
Drew and Flewelling defined two additional lines parallel to
the primary line. The first was at 15% of maximum density
(i.e. ratio of actual stand density to maximum density
attainable in a stand with the same biomass) which
approximated the density at initial crown closure. The
second was at 55% of maximum density, and it defined the
lower limit of what they termed the "imminent competition-
mortality" zone. Within this zone Drew and Flewelling
concluded that annual net growth, due to increased stand
mortality, would be considerably lower than that observed at
40 to 55% relative density. Thus, the 55% relative density
line appeared to define the previously considered optimum
area limit, i.e. a leaf area limit above which as Waring (55)
indicated, net production declined due to increased
mortality.

Application of leaf area manipulation and the
self-thinning rule to silvicultural management has been
previously discussed (9, 19). Actual application would
depend upon management objectives (21). However, it has been
generally concluded that for maximization of total wood
production, stands should be managed at LAIs or stand
densities approaching the line of imminent mortality (40-55%
of maximum density). If larger diameter trees are desired,
the stand should be managed at leaf areas closer to that of
initial crown closure (15-40% of maximum density). The two
zones of density or LAI management correspond to the
previously considered generalization (55), i.e. maximum
production per unit leaf area occurs at relatively low LAI's
after crown closure, while maximum biomass production occurs
at high LAI levels where production per unit leaf area is not
maximized.

3.3.2 Manipulation of Soil Physical Properties

Cultural treatments which seek to minimize drought stress
of the seedling or established stand are designed to increase
the supply of available soil moisture and/or enhance the
trees ability to obtain available reserves. The former is
usually accomplished by some type of physical soil
manipulation while the latter may also be enhanced by
fertilization, understory control and thinning (Section 3.3.3
and 3.3.4).

Manipulation of soil physical properties is usually done
during site preparation for planting. Treatments are
directed at ameliorating harvesting's detrimental impacts
and/or improving the site's inherent productivity. The
primary harvesting impacts on water relations of the future
stand are through soil disturbance or compaction and
destruction or disturbance of the forest floor. Compaction
causes increased soil bulk density and soil strength,
decreased soil aeration, and altered soil water transfer.
A 10% increase in soil bulk density commonly occurs after

equipment has passed over a point five or less times and
may result in marked growth declines for a number of years
after planting (11, 46, 48). A primary modifier of
compaction's significance to stand growth is the area
actually compacted which can vary from 10 to 40+%. In stands
influenced by relatively "normal" harvesting, skid-trail
compaction has been estimated to account for a 10 to 17%
decrease in total stand volume (48).

In addition to the above, an unavoidable side affect of
many silvicultural treatments is disturbance or destruction
of the forest floor. This decreases the insolating and
mulching effects of the forest floor and thus, may significantly increase drought-stress in both young and closed
canopy stands (13, 34, 38). Drum chopping and some disking
treatments have been credited with creating artificial
mulches and thereby performing or enhancing the role of the
forest floor and increasing seedling survival (3).

As in ameliorative treatments, most practices designed to
enhance a site's inherent water-limited productivity, with
the exception of herbicides, involve some form of tillage -
the most common types being disking and ripping/subsoiling
(13, 38, 48). On heavy to medium textured soils, disking
generally improves soil structure and aeration of the surface
soil. On drought-prone sites, incorporation of organic
matter increases soil moisture retention capacity (5).
Seedling root systems on heavy-textured disked sites exhibit
more diffuse root systems which are not as strongly oriented
along the planting slit (18).

A primary limitation of disking is its ability to
influence subsoil properties. Depending on the nature of the
soil, ripping in addition to disrupting surface compaction
influences site productivity in two ways: (1) it allows
vertical and lateral root penetration along ripping fractures
and 2) on rocky shallow soils, it enhances root microsites
through the accumulation of soil fines in the ripping slit.
On fine textured soils ripping may break traffic or

plow-pans, shallow hardpans or fragipans and thus increase
rooting volume, internal soil aeration, and drainage (35).
In reviewing research in which sites had been disked or
ripped, Froehlich and McNabb (11) reported changes in
survival of -9 to +39% and increases in height growth of 6 to
73%.

As a second primary function, site preparation limits
competition for limited soil moisture from non-crop
vegetation for limited soil moisture (41). The use of
herbicides and mechanical control treatments have become
an established part of many forest operations, especially on
drought-prone sites (7, 14).

3.3.3 Fertilization

Fertilization has normally been used to promote survival
and/or growth where specific nutrient deficiencies are the
primary limiting factors (38). Fertilization has not often
been practiced with the goal of increasing tree drought
resistance and in fact may increase competition for limited
moisture reserves. However, common fertilizer nutrients are
also known to be critical in the plant's resistance to
drought-stress. Nutrients such as potassium (K), phosphorus
(P) and nitrogen (N) have been shown to alter physiological,
water-related responses in a variety of plants (39, 40)
including forest species (30). Nutritional deficiencies in
P and N appear to first decrease root hydraulic conductivity
and subsequently, epidermal (stomatal) conductance (40).
Furthermore, these decreases appear to be related to membrane
fluidity and/or permeability (39, 40). Potassium is believed
to be an osmotic agent in the opening and closing of stomata
(30). For these reasons, well planned strategic
fertilization, especially in the nursery, may be directly
critical for development of seedling drought resistance.

Fertilization of more mature closed canopy stands which
are near their maximum LAI for their sites has often proven
ineffective in increasing growth of stem wood even where

drought-stress was not a major limiting factor (1, 2, 38). Fertilization of these lands has generally been most effective when applied after the stand has been thinned. It is thought that the reduction in LAI caused by thinning allows more light and growing space for remaining trees and often decreases the competition for soil moisture reserves. The roles of nutrients in the conservation and utilization of water under these conditions have not been fully studied with forest species. While fertilization has increased stand growth in relatively arid environments, it is not known if response to fertilization has been due to a direct nutritional enhancement of stand drought resistance.

3.3.4   Thinning and Understory Control

In Section 3.3.1 the relationship between leaf area, stand density, and non-crop or understory leaf-area was considered. These relationships are the basis for thinning and understory control. Thinning of plantations may be done for a variety of economic and/or management objectives (22). However, from a drought-stress/biological productivity perspective, thinning regimes have a common basis. When a closed canopy stand is thinned, its leaf area is at least for a time reduced. The immediate influence of leaf area reduction is generally thought to be a decrease, in stem-wood production (23). However, if the stand is in the "zone of imminent mortality" (8, 9) (Figure 1), thinning will decrease leaf area and may shift LAI back toward the optimum level, thus minimizing loss in stand wood production. In fact, thinning in this zone may be a harvest of what would be the stand's "mortality".

The loss in stand wood production depends on the severity of thinning and the time needed to re-occupy the site. Thinning may increase available soil moisture reserves and thereby temporarily enhance growth of remaining trees. However, other factors are also significant; physiological age or maturity, and stand condition may also be critical

(22). The stand's ability to rapidly respond to treatment
is critical if non-crop or understory vegetation is a major
competitor. Normally within two to three years of thinning,
leaf area of rapidly expanding ground vegetation may consume
much of the thinning-related-increases in available reserves.
On drought-prone sites, control of ground vegetation is often
critical to thinning responses (23).

3.4 INTRASPECIFIC GENETIC VARIATION IN TREE PHYSIOLOGICAL RESPONSES TO DROUGHT-STRESS

Before forest stands can be managed from a genetic
perspective for resistance to drought-stress, it is first
critical to identify those characteristics of drought
resistance which are under genetic influence (see Table 1).

Differences in survival are the first whole plant
manifestation of differences in drought resistance and
are an important driving force for current research in the
area of genotypic selection for resistance. Field tests gave
the first important indication of the presence of genetic
differences in survival (69), but they are unreliable because
natural drought periods cannot yet be predicted with
sufficient accuracy. Most of the research emphasis has been
on survival during the first two years.

Evidence is accumulating however that genetic differences
in survival at more mature ages also exist among provenances
of Pinus taeda plantations located in drought-prone areas
on the western edge of its range; seed sources orginating
east of the Mississippi River show more mortality than those
originating on the west side (31, McKinley and van Buijtenen,
unpublished) (Table 2). However, it should be recognized
that factors other than water also influence survival.

Intraspecific variation in epidermal conductance has been
widely demonstrated in a large number of forest species
including Pinus sp. (4, 29, 44), Populus sp. (26), and
Pseudotsuga menziesii (10, 51). In Pinus taeda, xeric
sources possessed fewer stomata per unit leaf surface area
than did mesic sources (28, 29, 50, 53). Transpiration rates

of two Pinus radiata clones differed by about one-half from
each other (42); the clone transpiring most rapidly had a
lower survival rate in the field (4). Transpiration is also
reduced by reduction in evaporative surfaces such as shoots
and leaves (23). Pinus sp. from xeric locations had slower
rates of shoot growth than those from mesic habitats (54, 57,
63, 64, 65).

Genetic variation in Pinus taeda root development has
been shown with deeper root penetration and more laterals
associated with drought-resistant genotypes (53). Recent
studies show that seedlings of drought-resistant families
produce a larger number of small roots during root
regeneration after outplanting (van Buijtenen, unpublished).

Intraspecific variation in osmotic adjustment capacity
has been demonstrated in Pinus taeda seedlings (20);
seedlings from the western edge of the southern pine region
did not osmotically adjust as well as those from the eastern
side when subjected to severe drought-stress (44). Sources
of Juglans nigra differed in their capacity to osmotically
adjust both in leaves and roots (37). Seedlings of an
Ontario, Canada source exhibited increased solute content,
whereas a New York source showed no evidence of osmotic
adjustment. More research is required to establish the
heritability of this physiological mechanism of drought
resistance in other commercial wood species.

Parker and Pallardy (37) observed differences in tissue
elasticity between sources of Juglans nigra in both roots
and shoots utilizing the pressure-volume technique. Refine-
ment of techniques for measuring elasticity and hydraulic
conductivity will undoubtedly allow for more demonstrations
of intraspecific variation in the near future.

There is evidence that cuticular waxes play a role in
drought resistance in several forest species. Wax is found
in all the stomatal cavities of Pinus taeda pine, some of
which are totally sealed, while others only partially.
Genetic differences have not yet been shown (van Buijtenen

unpublished), but P. taeda will respond strongly to water deficits by forming greater amounts of wax in the stomatal cavity. Desiccation tolerance differences were observed in P. taeda with xeric sources having a lower moisture content in dead needles than do mesic sources (53).

3.5 INTERACTION OF GENETICS AND SILVICULTURE IN DROUGHT STRESS-MANAGEMENT

3.5.1 Selecting for Drought Resistance

Selecting for increased survival when seedlings or trees are subjected to water deficits is important because whole plants utilize a variety of resistance mechanisms to resist drought-stress. Testing has been done under a wide range of conditions including field tests, the use of plastic shelters, greenhouses, and growth chambers. Field tests are often unreliable because of variability in climate and inability to accurately predict drought occurrences. Plastic shelters have been used successfully (52), but they showed great variability in the rate at which the drought-stress developed.

Greenhouse and growth-chamber tests have also been used for drought resistance screening. Their success depends on the drying method used and the rate at which stress develops. Two approaches can be used: watering the seedlings thoroughly and letting them dry down until a predetermined degree of mortality has occurred, or subjecting the trees to a predetermined amount of stress and observing the resulting mortality and/or physiological response. The first method has been used successfully in duplicating survival differences observed in field results. The second method is most useful for observing physiological responses. It is very difficult to control stress levels accurately enough for a sustained period of time to relate drought-stress to genetic differences in survival. Most of the research performed with these methods evaluated drought resistance during the first two years of growth.

3.5.2 Tissue Culture

Tissue culture has been used to investigate drought resistance in tissues with low water potential (36). Tissue culture compliments and offers some advantages over research based on whole plants. First, drought resistance mechanisms such as increased rooting, decreased stomatal conductance and leaf drop can be segregated so cellular tolerance mechanisms of drought resistance at low water potential can be investigated (Table 1); and, second, a large number of different germplasm sources can be screened for drought tolerance at low water potential in a short period of time and in a small space. Therefore, screening genotypes for drought resistance could greatly aid in accelerating existing tree improvement programs. Preliminary results suggest that tissue culture techniques have applicability for predicting growth potential in the field under stressed and nonstressed conditions (36). However, drought-resistant plantlets derived from tissue culture have not yet been tested in the field. A proposed scenario for integrating tissue culture into a conventional tree improvement program is indicated in Figure 2 and discussed more fully in other references (25, 36).

3.5.3 Matching Species with Site

Choosing the proper species is the most powerful genetic method we have of managing drought-stress. Species differ greatly in their site adaptability. Some species are very broadly adaptable and can be grown on a very wide range of sites, while other species are highly specialized. In the first category we find many of the pioneering species, such as Pinus taeda, Pseudosuga menziesii, and Pinus strobus which are used extensively for planting. Other species which appear to have more narrow site requirements such as Fraxinus pennsylvanica and Taxodium distichum, which naturally occur on rather wet sites, often do well on a wide range of sites

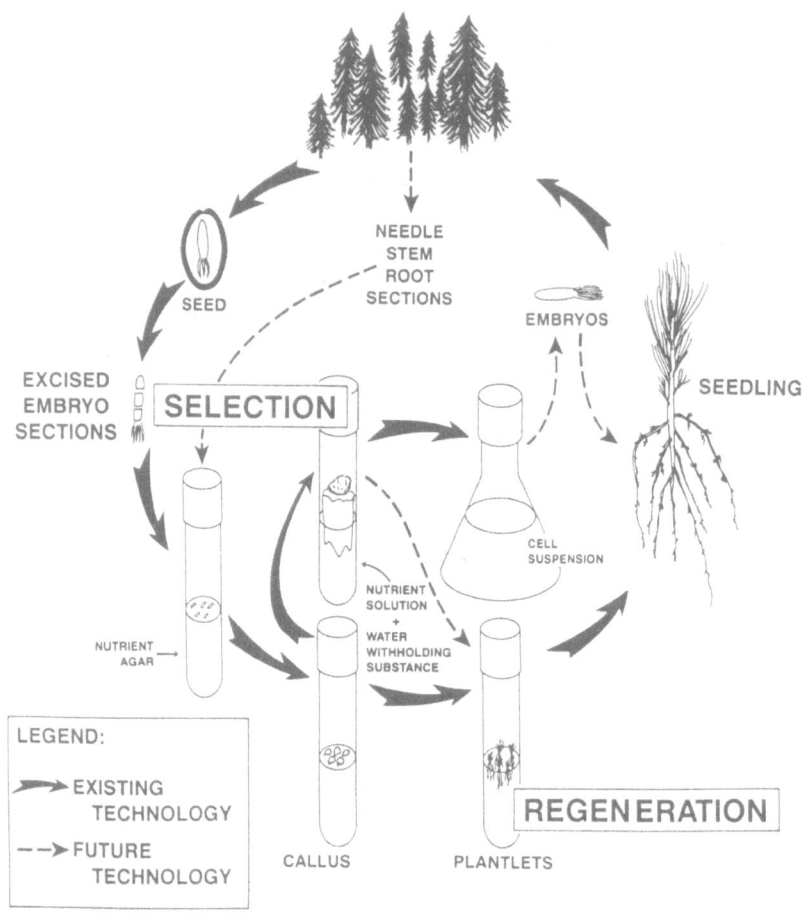

FIGURE 2.   Proposed scenario for the utilization of tissue
            culture in evaluation for drought-tolerance in
            pine.

when planted and properly managed (6, 47).   The reactions of
a species to a variety of sites are not predictable from

performance on its natural site. Similarly the use of exotic
trees in other locations is highly unpredictable based on the
performance of the trees at their original location (68).
Who would have, for instance, been able to predict the
tremendous success of Pinus radiata in Australia, based on
its performance in California? Other genera such as Populus
sp. and Platanus sp. have a very narrow range of sites to
which they are adapted. These are very important
considerations in trying to match the species to the site,
and a good knowledge both of the soils and of silviculture is
needed to make such a match.

3.5.4 Matching Species with Site

The first consideration in planting is the choice of the
proper geographic seed source (58). For instance,
Weyerhaeuser Co. plants Pinus taeda seed originating from
North Carolina on sites in Arkansas (31). They are however,
very careful to plant them only on the best sites. On the
more drought-stressed sites they plant the native Arkansas
sources which are adapted to these sites. Yeiser et al. (66)
showed that southern Louisiana sources may be well adapted to
high site index areas in northern Mississippi, but should not
be planted in southeast Texas. Seed sources from east of the
Mississippi River show higher mortality in plantations at the
western edge of the P. taeda range (Table 2).

How to conduct a breeding program for specific sites is a
difficult question which is highly influenced by economics.
Usually such sites are limited in acreage, which makes it
uneconomical to undertake an intensive breeding program. In
such cases, however, it is possible to test individual
families on extreme sites, collect seed separately by family
in the seed orchard, raise the seedlings by family in the
nursery, and plant the seedlings from families adapted to the
extreme sites back to those same sites. Some of the genetic

TABLE 2. Trees dying (% of planted) at ages 6-10 for eastern and western sources of <u>Pinus</u> <u>taeda</u> L. at three planting locations. (McKinley and van Buijtenen, unpublished).

	Planting Location		
	Robertson Co. (TX)	Trinity Co. (TX)	Sabine Par. (LA)
Eastern Sources			
South Carolina	9.4	5.9	6.6
Florida	11.0	29.2	2.5
Georgia - 1	4.5	*	4.7
Georgia - 2	14.2	*	1.2
Louisiana - 1 (Evangeline Parish)	3.0	0.0	1.1
Louisiana - 2 (Livingston Parish)	3.7	13.7	4.5
Western Sources			
Arkansas	0.8	1.0	1.2
Texas-1 (Fayette Co.)	0.0	0.8	1.1
Texas-2 (Fayette Co.)	1.1	1.5	0.0
Texas-3 (Bastrop Co.)	0.0	2.3	2.2
Texas-4 (Bastrop Co.)	0.0	1.5	3.4
Texas-5 (Bastrop Co.)	0.0	4.4	3.1

* Source not planted

gain is lost because there is no control of pollen source, but there is also a tremendous reduction in cost, making this the method of choice for small acreages. If large acreages are available, then the standard seed orchard approach is quite feasible.

Another approach which may become practical in the future is the development of clonal lines for specific sites. The degree of adaptation could be increased tremendously while clonal propagation initially might work better for a more limited acreage because of cost and limited availability. Tissue culture will become more important in the future as a means of clonal propagation (25) (see Figure 2, Section 3.5.2).

3.5.5 Cultural Treatment

Cultural treatments have a greater effect and quicker results than genetic improvement, but may cost more. The best means of managing drought-stress is by a combination of both genetics and silviculture. This is very much like the experience in agriculture over the last century. Yields have improved greatly, not only by breeding of more productive strains, but also by the development of more intensive agricultural methods. On dry sites one should plant a drought-resistant, geographic source of a "given" species. This should be supplemented with proper silvicultural methods.

There is a paradox in this approach, however. Intensive silviculture and the use of the best genetic material will give the highest return on the investment on the best sites. On the other hand, to grow a tree at all on poor sites requires more intensive management. As a consequence there are economic constraints on what is possible. When one moves from management on a good site to a poor site the economic returns become marginal.

3.5.6 Irrigation Studies

Optimization of soil moisture with irrigation may increase yields by 20-70% (14, 27, 45, 59). Yet, despite increases in yield, it is generally uneconomical to irrigate in most forest situations. Periodic irrigation resulting in preconditioning to drought-stress increased pine root regeneration (20). This may be an important procedure for insuring stand survival on drought-prone sites.

Jarvis (23) points out that experimentally we have a problem when we try to consider the effect of drought-stress on the growth of stands in the field. There have been very few experiments in which water is the independent variable. Jarvis (23) suggests that irrigation experiments to determine the influence of water availability on forest stand productivity be implemented. He recommends that the main

treatment be application of water at different rates with
additional treatments of thinning and fertilizer. The data
would then be used to develop a management-oriented,
process-based model of growth and water use of forest
plantations.

3.6 SUMMARY

Genetic and silvicultural management techniques increase
forest stand productivity during drought-stress by modifying
tree physiology. Site improvement generally enhances
water availability to the stand and increases or facilitates
root growth. Fertilizers enhance forest nutrient status
and modify its water relations presumably by altering tree
cell membrane permeability and cellular, osmotic properties.
It is not always clear whether fertilization enhances or
decreases tree drought resistance. Forest stand evapotranspiration is directly related to stand leaf area.
A balance between transpiration and forest stand productivity
can be accomplished with silvicultural practices whereby crop
leaf area is optimized by eliminating non-crop leaf area and
thinning the stand. The identification of physiological
mechanisms of drought resistance can aid the tree breeder in
selection processes. Intraspecific variations in epidermal
conductance and rooting in tree species have been well
documented, but the heritabilities of osmotic adjustment,
elasticity and hydraulic conductivity have been demonstrated
in only a few species. Tissue culture may have utility in
preliminary, screening tests for drought resistance as well
as providing methodology for clonal propagation for
drought-prone sites. Maximization of response to and/or
feasibility of intensive cultural treatments must be based
upon development of genetic materials which can tolerate
and/or respond to the cultural treatments under given site
conditions.

REFERENCES

1. Allen, H.L. 1983. Forest fertilization. pp. 129-160 in Allen, H.L. (compiler). North Carolina State Forest Fertilization Cooperative Forest Soils Short- course. School of Forest Resources, North Carolina State University, Raleigh.

2. Allen, H.L. and R. Ballard. 1983. Fertilization of loblolly pine. pp. 163-181 in Kellison R.C. (ed.). Symposium on the Loblolly Pine Ecosystem (East Region). School of Forest Resources, North Carolina State University, Raleigh.

3. Bennett, K.J. and D.A. Rook. 1978. Stomatal and mesophyll resistances in two clones of Pinus radiata D. Don. known to differ in transpiration and survival rate. Aust. J. Plant Physiol. 5: 231-238.

4. Bilan, M.V., C.T. Hagan and H.B. Carter. 1977. Stomatal opening, transpiration, and needle moisture in loblolly pine seedlings from two Texas seed sources. For. Sci. 23: 457-462.

5. Burns R.M. and E.A. Hebb. 1972. Site preparation and reforestation of droughty acid sands. USDA Agricultural Handbook No. 426, 61 p.

6. Byram, T.D., W.J. Lowe, C.R. McKinley, J.F. Robinson, A.F. Stauder, and J.P. van Buijtenen. Thirty-First Progress Report of the Cooperative Forest Tree Improvement Program. Circular 266, Texas Forest Service, College Station.

7. Carter, G.A., J.H. Miller, D.E. Davis and R.M. Patterson. 1984. Effect of vegetative competition on the moisture and nutrient status of loblolly pine. Can. J. For. Res. 14: 1-9.

8. Drew, T.J. and J.W. Flewelling. 1977. Some recent Japanese theories of yield density relationships and their application to monterey pine plantations. For. Sci. 23: 517-534.

9. Drew, T.J. and J.W. Flewelling. 1979. Stand density management: an alternative approach and its application to Douglas-fir plantations. For. Sci. 25: 518-532.

10. Ferrel, W.K. and S. Woodward. 1965. Effect of seed origin on drought resistance in Douglas-fir. Ecology. 47: 499-503.

11. Froehlich, H.A. and D.H. McNabb. 1984. Minimizing soil compaction in Pacific Northwest forests. pp. 159-192 in Stone, E.L. (ed.) Forest Soils and Treatment Impacts. Department of Forestry, Wildlife and Fisheries, University of Tennessee, Knoxville.

12. Gholz, H.L. 1982. Environmental limits on aboveground net primary production, leaf area, and biomass in vegetation zones of the Pacific northwest. Ecology 63: 469-481.

13. Ginter, D.L., K.W. McLeod and C. Serrad, Jr. 1979.
    Water stress in longleaf pine induced by litter
    removal. For. Ecol. and Mgt. 2: 13-20.
14. Gjerstad, D.H., L.R. Nelson, and P.J. Minogue. 1984.
    Chemical forest vegetation management. pp. 175-184
    in Karr, B.L., J.B. Baker, and T. Monaghan
    (eds.). Proceedings of the Symposium on the Loblolly
    Pine Ecosystem (West Region). Mississippi Cooperative
    Extension Service, Mississippi State, MS.
15. Gordon, J., P. Farnum and R. Timmis. 1983.
    Theoretical maximum phytomass yields as a guide
    for yield improvement. pp. 19-34 in Thieleges,
    B.A. (ed.). Proceedings of Seventh North American
    Forest Biology Workshop: Physiology and Genetics of
    Intensive Culture. Dept. of Forestry, University of
    Kentucky, Lexington.
16. Gresham, C.A. and T.M. Williams. 1978. Water table
    depth and growth of young loblolly pine. pp. 371-376
    in Balmer, W.E. (ed.) Proceedings: Soil Moisture....
    Site Productivity Symposium. USDA, Forest Service,
    Southeastern Area, State and Private Forestry.
17. Grier, C.C. and S.W. Running. 1977. Leaf area of
    mature northwestern coniferous forests: relation to
    site water balance. Ecology 58: 893-899.
18. Haines, L.W. and S.G. Haines. 1978. Site preparation
    for regeneration. pp. 176-195 in Balmer, W.E. (ed.)
    Proceedings: Soil Moisture...Site Productivity
    Symposium. USDA, Forest Service, Southeastern Area,
    State and Private Forestry.
19. Harms, W.R. 1984. Applying silvics to stand
    management. pp. 33-40 in Karr, B.L., J.B. Baker and
    T. Monaghan (eds.). Proceedings of the Symposium on
    the Loblolly Pine Ecosystem (West Region). Mississippi
    Cooperative Extension Service, Mississippi State, MS.
20. Hennessey, T. C. and P.M. Dougherty. 1984. Charac-
    terization of the internal water relations of loblolly
    pine seedlings in response to nursery cultural
    treatments: Implications for reforestation success.
    pp. 225-243 in M.L. Duryea and G.N. Brown (eds.).
    Seedling Physiology and Reforestation Success.
    Martinus Nijhoff/Dr. W. Junk, Publishers. Boston, USA.
21. Hughes, J.H. and J.W. Herschelman. 1984. Density
    regulation--the plantation thinning equation.
    pp. 155-174 in Karr, B.L., J.B. Baker, T. Monoghan
    (eds.). Proceedings of the Symposium on the Loblolly
    Pine Ecosystem (West Region). Mississippi Cooperative
    Extension Service, Mississippi State, MS.
22. Hughes, J.H. and R.C. Kellison. 1983. Stocking
    control--silvicultural thinning during the rapid
    growth years. pp. 136-145 in Kellison, R.C. (ed.).
    Symposium on the Loblolly Pine Ecosystem (East
    Region). School of Forest Resources, North Carolina
    State University, Raleigh.

23. Jarvis, P.G. 1985. Increasing productivity and value of temperate coniferous forest by manipulating site water balance. pp. 39-74 in Ballard, R. (ed.). Weyerhaeuser Science Symposium, Forest Potentials - Productivity and Value, Weyerhaeuser Co., Centralia, WA.

24. Jones, M.M., N.C. Turner, and C.B. Osmond. 1981. Mechanisms of Drought Resistance in Plants. pp. 15-37 in Paleg, L.G. and D. Aspinall (eds.). The Physiology and Biochemistry of Drought Resistance in Plants. Academic Press, New York.

25. Karnosky, D.F. 1981. Potential for forest tree improvement via tissue culture. BioScience 31: 114-120.

26. Kelliher, F.M. and C.G. Tauer. 1980. Stomatal resistance and growth of drought-stressed eastern cottonwood from a wet and dry site. Silv. Gen. 29: 166-171.

27. Klawitter, R.A. 1978. Growing of pine on wet sites in the southeastern coastal plain. pp. 49-62 in Balmer W.E. (ed.). Proceedings: Soil Moisture...Site Productivity Symposium. USDA, Forest Service, Southeastern Area, State and Private Forestry.

28. Knauf, T.A. and M.V. Bilan. 1974. Needle variation in loblolly pine from mesic and xeric seed sources. For. Sci. 20: 89-90.

29. Knauf, T.A. and M.V. Bilan. 1977. Cotyledon and primary needle variation in loblolly pine from mesic and xeric seed sources. For. Sci. 23: 33-36.

30. Kramer, P.J. and T. T. Kozlowski. 1979. Physiology of Woody Plants. Academic Press, New York.

31. Lambeth, C.C., R.B. McCullough and O.O. Wells. 1984. Seed source movement and tree improvement in the Western Gulf region. pp. 71-86 in Karr, B.L., J.B. Baker, T. Monoghan (eds.). Proceedings of the Symposium on the Loblolly Pine Ecosystem (West Region). Mississippi Cooperative Extension Service, Mississippi State, MS.

32. Levitt, J. 1980. Responses of plants to environmental stresses. Academic Press, New York.

33. Linsley, R.K., M.A. Kohler and J.L.H. Paulhus. 1959. Applied Hydrology. McGraw-Hill, New York.

34. McLeod, K.W., C. Sherrod, Jr., and T.E. Porch. 1979. Response of longleaf pine plantations to litter removal. For. Ecol. and Mgt. 2: 1-12.

35. Nambiar, E.K.S. 1983. Root development and configuration in intensively managed radiata plantations. Plant Soil 71:37-47.

36. Newton, R.J., S. Sen and J. P. van Buijtenen. 1985. Growth changes in loblolly pine (Pinus taeda L.) cell cultures in response to drought-stress. pp. 64-73 in Proceedings of the 18th Forest Tree Improvement Conference. University of So. Mississippi, Long Beach, May 21-23.

37. Parker, W.C. and S.G. Pallardy. 1985. Genotypic variation in tissue water relations of leaves and roots of black walnut (_Juglans nigra_) seedlings. Physiol. Plant. 64: 105-110.
38. Pritchett, W.L. 1979. Properties and Management of Forest Soils. John Wiley and Sons, New York.
39. Radin, J.W. and J.S. Boyer. 1982. Control of leaf expansion by nitrogen nutrition in sunflower plants. Role of hydraulic conductivity and turgor. Plant Physiol. 69: 771-775.
40. Radin, J.N. and M.P. Eidendock. 1984. Hydraulic conductance as a factor limiting leaf expansion of phosphorus-deficient cotton plants. Plant Physiol. 75: 372-377.
41. Radosevich, S.R. 1984. Interference between manzanita (_Arctostaphylos patula_). pp. 259-270 _in_ pp. 259-270. M.L. Duryea and G.N. Brown (eds.). Seedling Physiology and Reforestation Success. Martinus Nijhoff/Dr. W. Junk, Publishers, Boston.
42. Rook, D.A. and J.F. Hobbs. 1976. Soil temperatures and growth of rooted cutting of radiata pine. N. Z. J. For. Sci. 5: 296-305.
43. Shroeder, P.E., B. McCandish, R. W. Waring and D. A. Perry. 1982. The relationship of maximum canopy leaf area to forest growth in eastern Washington. Nthw. Sci. 56: 121-130.
44. Seiler, J. R. 1984. Physiological response of loblolly pine seedlings to moisture-stress conditioning and their subsequent performance during water stress. Ph.D. Dissertation, Virginia Polytechnic Institute and State University, Blacksburg, VA.
45. Shoulders, E. 1974. Fertilization and Soil Moisture Management in Southern Pine Plantations. pp. 55-64 _in_ Proceedings: Symposium on Management of Young Pines. USDA, Forest Service, Southeastern Area, State and Private Forestry.
46. Simmons, G.L. and A.W. Ezell. 1983. Root development of loblolly pine seedlings in compact soils. pp. 26-29 _in_ Jones, E.P. Jr. (ed.) Proceedings of the Second Biennial Southern Silvicultural Research Conference. USDA Forest Service, Southeastern Forest Experiment Station, General Technical Report, SE-24.
47. Stauder, A.F. and W.J. Lowe. 1983. Geographic variability in growth of ten-year-old green ash families with East Texas. pp. 227-233 _in_ Proc. 17th Southern Forest Tree Improvement Conference, Athens, GA.

48. Switzer, G.L., D.M. Moehring and T.A. Terry. 1978. Clearcutting vs. alternative timber harvesting-stand regeneration systems: effects on soils and environment of the south. pp. 477-515 in Youngberg, C.T. (ed.). Forest Soils and Land Use. Department of Forest and Wood Sciences, Colorado State University, Fort Collins, Co. pp. 477-515.

49. Tadaki, Y. and T. Shidei. 1959. Studies on the competition of forest trees. II. The thinning experiment on small model stand of Sugi (Crytomeria japonica) seedlings. Nippon Rin Gakkaishi 41: 341-349.

50. Thames, J.L. 1963. Needle variation in loblolly pine from four geographic sources. Ecology 44: 168-169.

51. Untersheutz, P., W.F. Ruetz, R.R. Geppert and W.K. Ferrell. 1974. The effect of age, preconditioning, and water-stress on the transpiration rates of Douglas-fir (Pseudotsuga menziesii) seedlings of several ecotypes. Physiol. Plant. 32: 21-221.

52. van Buijtenen, J.P. 1966. Testing loblolly pine for drought resistance. Tech. Pap. No. 13, Texas Forest Service, College Station, TX.

53. van Buijtenen, J.P., M.V. Bilan and R.H. Zimmerman. 1976. Morpho-physiological characteristics related to drought resistance in Pinus taeda L. pp. 349-359 in Cannell, M.G.R. and F.T. Last (eds.). Tree Physiology and Yield Improvement. Academic Press, New York.

54. Venator, C.R. 1976. Natural selection for drought resistance in Pinus caribaea Morelet. Turrialba 26: 381-387.

55. Waring, R.H. 1983. Estimating forest growth and efficiency in relation to canopy leaf area. Adv. in Ecol. Res. 13: 327-381.

56. Waring, R.H., W.H. Emmingham, H.L. Gholz and C.C. Grier. 1978. Variation in maximum leaf area of coniferous forests in Oregon and its ecological significance. For. Sci. 24: 131-140.

57. Wells, C.O. and P.C. Wakeley. 1966. Geographic variation in survival, growth, and fusiform rust infection of planted loblolly pine. For. Sci. Monogr. No. 11, 40 pp.

58. Wells, O.O. 1969. Results of the southwide pine seed source study through 1968-1969. pp. 117-129 in Proc. of the Tenth Southern Forest Tree Improvement Conference, Houston, TX.

59. Wells, C.G. and D.M. Crutchfield. 1974. Intensive culture of young loblolly pine. pp. 212-228 in Proceedings: Symposium on Management of Young Pines. USDA, Forest Service, Southeastern Area, State and Private Forestry.

60. Westoby, M. 1977. Self-thinning driven by leaf area not by weight. Nature 265: 330-331.

61. Westoby, M. 1984. The self thinning role. Advances in Ecological Research 14: 167-220.

62. Williston, H.L. 1972. The question of adequate stocking. Tree Planters' Notes. 23(1) 2 pp.

63. Woesner, R.A. 1972. Crossing among loblolly pines indigenous to different areas as a means of genetic improvement. Silvae Genet. 21: 35-39.

64. Woesner, R.A. 1972. Growth patterns of one-year-old loblolly pine seed sources and inter-provenance crosses under contrasting edaphic conditions. For. Sci. 18: 205-210.

65. Wright, J.W. and W.T. Bull. 1963. Geographic variation in Scotch pine: results of a 3-year Michigan study. Silv. Gen. 12: 1-25.

66. Yeiser, J.L., J.P. van Buijtenen and W.J. Lowe. 1981. Genotype x environment interaction and seed movements for loblolly pine in the Western Gulf region. Silv. Gen. 30: 196-200.

67. Zahner, R. 1968. Water deficits and growth of trees. pp. 191-254 in Kozlowski, T.T. (ed.). Water Deficits and Plant Growth, Vol. II. Academic Press, New York.

68. Zobel, B.J. and J. Talbert. 1986. Applied Forest Tree Improvement. Wiley and Sons, N.Y.

69. Zobel, B.J. and R.E. Goddard. 1955. Preliminary results on tests of drought-hardy strains of loblolly pine (Pinus taeda L.). Res. Note, No. 14., Texas Forest Service, College Station, TX.

4. NUTRIENTS: USE OF FOREST FERTILIZATION AND NUTRIENT EFFI-
CIENT GENOTYPES TO MANAGE NUTRIENT STRESS IN CONIFER STANDS

J. L. Troth, R. G. Campbell, and H. L. Allen

Manager, Weyerhaeuser Company Forest Research Field Station,
Mountain Pine, AR 71956
Forest Soils Specialist, Weyerhaeuser Company Forest Research
Field Station, New Bern, NC 28560
Director, North Carolina State University Forest Nutrition
Cooperative, School of Forest Resources, Raleigh, NC 27695

ABSTRACT

Growth responses to fertilizers by many tree species
across a wide range of stand and site conditions indicate that
suboptimal nutrition frequently limits forest productivity.
Much of the increase in stemwood growth following fertilizer
additions is attributable to increased foliage production.
Other mechanisms which may contribute to accelerated stem
growth include increased productivity per unit of leaf area
and altered dry matter partitioning among tree components.
Several studies indicate that fertilization also improves
tree-soil water relations. Even within a single region and
species, the magnitude of fertilizer response varies greatly
among stands and sites. Existing procedures for evaluating
the nutritional status of forest trees and predicting responses to fertilizers leave much of this variation unexplained. Tree growth under low levels of soil nutrient availability and tree responses to fertilizers also vary among
genotypes; however, mechanisms responsible for genotypic
variation have not been identified. Genotypes with high
rooting density may be best suited for uptake of poorly mobile
ions, such as ammonium and phosphate, from soils with low
nutrient availability.

4.1 INTRODUCTION

Nutrient availability to forest stands can be readily increased through fertilizer amendments. However, it has only been within the last couple of decades that forest fertilization has become an established silvicultural practice applied on a large scale. At present, forest fertilization in North America is prescribed primarily for coniferous species, usually on intensively managed plantations of loblolly (<u>Pinus</u> <u>taeda</u> L.) and slash pines (<u>Pinus</u> <u>elliottii</u> Engelm.) in the southern United States and on both planted and natural stands of Douglas-fir (<u>Pseudotsuga</u> <u>menziesii</u> (Mirb.) Franco) in the Pacific Northwest. Operational fertilization of southern pine plantations can be separated into two broad categories: 1) applications of phosphorus (P) or nitrogen (N) plus P at approximately the time of planting on very poorly to somewhat poorly drained, acid soils of the Lower Coastal Plain and 2) applications of N or N plus P to established plantations throughout the southern pine region. Operational fertilizer additions to Douglas-fir are limited to N applications in established stands. Trial applications of other elements have not produced widespread economical growth increases in these regions (2, 28, 34), although potassium and micronutrient additions show promise on some sites (28).

This paper will focus on the N and P nutrition of conifers and the effects of additions of these nutrients to coniferous stands. Discussion will emphasize, but not be limited to, the southern pines and Douglas-fir. The management practices considered will be forest fertilization and the potential use of nutrient efficient genotypes. Purely operational considerations such as fertilizer rates, materials, and timing of applications will not be discussed.

Specific objectives will be to 1) summarize the extent to which forest productivity is limited by nutrition, 2) review the mechanisms of tree and stand response to improved nutrition, 3) describe procedures used to characterize nutrient status of stands and predict fertilizer response, including limitations to use of these methods, 4) review opportunities to develop nutrient efficient genotypes, and 5) identify research needed to better identify and manage stands in which productivity is significantly limited by nutrient stress.

4.2 OVERVIEW OF GROWTH RESPONSE

Extensive series of field trials established in the 1960's and 1970's by industry-university forest fertilization research cooperatives provided measurements of fertilizer response for loblolly (2) and slash pines (28) in the south and Douglas-fir (71) in the northwest across a wide range of stand and soil conditions. These measurements demonstrated that nutrition commonly limits forest productivity, as indicated by increased tree growth following nutrient additions. However, it was also apparent that the magnitude of volume responses varied greatly among trials and that reliable diagnostic criteria for predicting individual stand responses were needed.

Growth responses to fertilization at or near the time of planting have been associated primarily with P additions to very poorly to somewhat poorly drained, acid soils in the southeastern Lower Coastal Plain (20, 32, 33, 76, 95). In one set of trials, addition of P to beds approximately doubled loblolly pine plantation volumes at ages 12 to 13 (Table 1). Dramatic increases in slash pine growth were still apparent 17 to 20 years after single P applications (75). Nitrogen sometimes produced an additional growth benefit when applied with P, but N alone was seldom beneficial at planting time (33, 76).

TABLE 1. Effect of 45 to 67 kg P/ha applied at planting time on 12- and 13-year-old loblolly pine plantations on bedded sites. Values are means (+ standard error) for 9 locations in the Lower Coastal Plain (calculated from data of Gent et al. 1986a).

Treatment	Trees/ha	Height	DBH	Volume
		m	cm	m^3/ha
Bed	1116	7.4	10.7	53
	(+46)	(+0.5)	(+0.7)	(+9)
Bed + Fert.	1163	10.2	14.4	103
	(+51)	(+0.3)	(+0.3)	(+7)

Growth responses of established stands to fertilization have been reported for many species across a wide range of stand and site conditions. In both the southern Upper Coastal Plain and Piedmont, many loblolly pine stands respond to fertilizers. In some stands, full response is achieved with N alone, but response to N+P frequently exceeds response to N. Additions of P alone generally produce substantial pine growth increases only on the P deficient soils of the Lower Coastal Plain, but even on those sites established stand applications of N+P often provide greater gains than additions of P alone (2, 31). In the Pacific Northwest both thinned and unthinned Douglas-fir stands respond strongly to N additions (Table 2). Within both the southern pines and Douglas-fir, fertilizer response varies widely among sites even when considering a single treatment within a single physiographic region (2, 72) or within groups of similar soils (49).

TABLE 2. Eight-year volume response to N fertilization in second-growth Douglas-fir stands (from data of Peterson and Gessel 1983)

	Average 8-Year Response			
	95 Unthinned Stands		35 Thinned Stands	
Treatment	Gross	Net	Gross	Net
	(% increase) m^3/ha			
224 kg N/ha	(13) 24.0	(11) 16.8	(20) 32.0	(20) 31.2
448 kg N/ha	(17) 32.0	(13) 20.0	(23) 37.6	(23) 36.8

4.3 RESPONSE MECHANISMS

Increased stem growth following fertilization has been hypothesized to result from both increased foliage production and greater net primary productivity per unit of foliage area. More recently, increased allocation of carbon to stemwood formation has been proposed as a possible mechanism of fertilizer response (4, 52). Several studies have also indicated that fertilization favorably alters soil-tree water relations (15, 42, 84). Few studies have included measurements of root growth responses or related these to changes in foliage production, aboveground growth, tree water status, or nutrient uptake.

4.3.1 Foliage production and productivity per unit of foliage

On responsive sites, fertilization has dramatically increased crown development and foliage production. In a 20-year-old Douglas-fir stand, N fertilization increased needle

length, needle width, number of needles per shoot, first-order
branch length, and number of second-order branches (19).
Needle surface areas (both sides) of control and fertilized
trees averaged 0.65 and 0.78 cm^2, respectively. Number of
needles was not increased in the first year after fertilization, since primordia were initiated prior to treatment; however, in subsequent years, fertilized trees produced almost
50% more needles per shoot than did control trees. Five years
after treatment, N fertilization had approximately doubled the
number of shoots in a main branch whorl of Douglas-fir (16).
First-year increases in individual needle or fascicle weights
following fertilization have also been reported for loblolly
pine (54, 94), jack pine (Pinus banksiana Lamb.) (91), balsam
fir (Abies balsamea (L.) Mill.) (83) and lodgepole pine (Pinus
contorta Dougl.) (92). Maki (54) reported that N fertilization increased needle length of loblolly pine and the proportion of fascicles with 4 or more, rather than 3, needles.
Fertilization may either increase (63), decrease (16), or have
no effect on needle longevity (58), with differences perhaps
due to both species and degree of nutrient stress prior to
treatment.

Seven years after treatment, total needle mass of co-
dominant trees in a 24-year-old Douglas-fir stand was increased 90% by either N fertilization or thinning and by 271%
when the two treatments were combined (16). Two years after
treatment, N fertilization of radiata pine (Pinus radiata D.
Don) increased dry weight ha^{-1} of 1-year-old needles by 80 and
122% in unthinned and thinned treatments, respectively (58).
Increased dry weight of needle litter fall beneath loblolly
pine stands was still apparent 5 years after 3 consecutive
annual fertilizations (54). In both Douglas-fir and radiata
pine, fertilization increased foliar mass primarily in the
upper crown of unthinned stands, but at all crown positions in
thinned stands (16, 58).

Brix (14) reported that N fertilization of Douglas-fir increased needle chlorophyll (a + b) concentrations and, in detached well-watered shoots, rates of both net photosynthesis and dark respiration per unit of leaf area. In August of the first season after treatment, net photosynthesis rates in current-year shoots from control and treated trees averaged 8.3 and 10.3 mg CO_2 dm^{-2} hr^{-1}, respectively. No significant effects on photosynthesis rates were detected in older shoots despite substantial increases in levels of foliar N and chlorophyll. The magnitude of increases in the rate of net photosynthesis were found to vary with both light intensity and foliar N concentrations. No significant treatment effects on net photosynthesis rates were apparent at light intensities up to 2,000 ft-c (14). At higher light intensities photosynthesis rates of treated trees exceeded those of controls with a maximum gain of 78% measured at 5,000 ft-c. In experiments where foliar N concentrations were varied widely by varying N application rates, photosynthesis rates were linearly related to foliar N concentrations between approximately 0.9 and 1.6% N (17). The apparent optimum foliar N concentration for photosynthesis was 1.7%.

Brix (18) evaluated the relative importance to aboveground dry matter production of increased foliar mass versus increased productivity per unit of foliar mass for a 7-year period following N fertilization and thinning of Douglas-fir. The majority of stemwood dry matter response to all treatments, 63 to 80%, was attributed to increased foliar mass, indicating that crown expansion plays a major role in fertilizer response. Increased productivity per unit of foliage was a major component of response only during the first three years after fertilization. In the latter years of the 7-year period, aboveground production per unit of leaf area or weight was actually lower in fertilized than in control trees. Nitrogen fertilization of Corsican pine (_Pinus_ _nigra_ var.

maritima (Ait) Melv.) also increased both needle area and net primary productivity per unit of leaf area (63). However, in a study of radiata pine response, N increased foliage production with no apparent effect on aboveground dry matter increment per unit of needle weight (58). Brix (18) postulated that stand response to fertilizers will depend on how deficient the stand is in foliage relative to that required for maximum production. Unfortunately, knowledge of what constitutes an optimum amount of foliage for stand production and of the interacting effects of nutrition and other limiting factors on foliage production are currently inadequate to utilize assessments of leaf mass or area to predict response to fertilizers.

4.3.2 Root response and carbon partitioning among tree components

Few studies in forest stands have attempted to quantify the effects of fertilizers on distribution of dry matter growth among all tree components, both above- and belowground, to determine the effects of improved nutrition on relative allocation of growth to stemwood. Sutton (82) stated that high fertility seems generally to reduce tree root growth in favor of shoot growth. Russell (79) reported that improved N supply reduced size of root systems relative to shoots of various agronomic species, although added N frequently increased root growth, but in reduced proportion relative to that of the shoot.

Ledig and Perry (51) pointed out that shoot/root ratios of tree seedlings also change with seedling size and concluded that reported differences among fertilizer treatments in shoot/root ratios were frequently attributable to differences in tree size, rather than altered relationships between relative rates of shoot and root growth. The relationship

between shoot and root dry weights of Douglas-fir was not
altered by fertilizers when substantial differences in tree
size due to treatments were taken into account (22). In many
reports it is not clear whether changes in shoot/root ratio
associated with differences in nutrition are true shifts in
relative growth or simply result from comparing trees of
different sizes or ages.

Axelsson (4) described the effects of annual fertilization on dry matter production and allocation among both above-
and belowground components of a 20-year-old Scots pine (Pinus
sylvestris L.) stand during the sixth year after initiation of
treatments. Allocation of dry matter production to fine roots
(< 2mm in diameter) was reduced from 39% in unfertilized
plots to 17% in those fertilized annually (Table 3). Percentage allocation of dry matter to all other tree components,
including coarse roots (> 2mm), was greater in annually fertilized than in control trees. Consideration of aboveground

TABLE 3. Total biomass production in 1979 and its distribution among tree components of a 20-year-old Scots pine stand
in which annual fertilization treatments were initiated in
1974 (adapted from Axelsson 1983)

| | | Distribution Among Components | | | | |
| | Total | | | Stem + | Roots | |
Treatment	Production	Needles	Branches	Stump	>2mm	<2mm
	kg/ha/yr			%		
Control	4,700	18.2	15.6	14.3	12.6	39.2
Annual Solid						
Fertilization	10,500	24.3	21.6	17.7	19.0	17.4

tree components only would have indicated small treatment effects on dry matter partitioning among needles, branches, and stems. Single N applications to Douglas-fir (18) and radiata pine (58) have increased allocation of aboveground dry matter production to branches and foliage, while reducing dry matter allocation to stemwood; however, no root measurements were included in these studies. In contrast, N fertilization did not alter biomass distribution among stems, branches, and foliage of a 3-year-old loblolly pine plantation 2 years after treatment, despite substantial increases in dry weight (5).

There is little information regarding the effects of single fertilizer applications to forest stands on root system growth and dry matter allocation to coarse and fine roots. However, a variety of effects of fertilizers on production and numbers of mycorrhizal root tips have been reported. Ammonium nitrate, applied with or without P, reduced the number of mycorrhizal root tips beneath a loblolly pine plantation during the first season after treatment, but not during the second season (60). Additions of P alone did not alter the number of mycorrhizal root tips. Beneath western hemlock (Tsuga heterophylla (Raf.) Sarg.), urea caused a decrease in some types of mycorrhizae and an increase in other types (35). In radiata pine, a complete fertilizer increased the proportion of root tips which were mycorrhizal on poor sites, but reduced the proportion of root tips which were mycorrhizal on high quality sites (81).

Grier et al. (39) observed that trees in older sub-alpine forests allocated substantially more carbon to root growth than did trees in younger stands. They hypothesized that reduced nutrient availability in the older stands contributed to increased carbon allocation to fine roots. Similarly, comparisons of dry matter production in two 40-year-old Douglas-fir stands on sites with high and low fertility revealed that, although total net primary production differed by

only 13% between the two stands, 8% of total net primary
production was in fine roots (<2 mm) on the more productive
site as compared to 36% on the poorer site (47). Conversely,
46% of net primary production occurred in the stemwood component on the better site as compared to 27% on the poorer site.
These results suggested that apparent differences in site
productivity based on aboveground measurements were partially
due to greater investment in fine roots on less fertile sites.
In contrast, other studies have suggested that fine root
turnover rates were higher in soils with high N mineralization
rates than in soils with low mineralization rates, so that
fine root production did not decrease, and might have increased, along a gradient of increasing soil N availability
(67). The latter authors hypothesized that higher fine root
biomass on poor sites was an adaptation to low nutrient
availability characterized by maintenance of long-lived, high
density fine root systems with low turnover rates.

4.3.3 Interactions with tree-soil water relations

Several studies have demonstrated potentially beneficial
changes in tree-soil water relations associated with fertilizer additions. Reduced water stress in fertilized trees as
compared to unfertilized controls, despite reduced soil water
availability or content beneath fertilized plots, has been
observed in stands of Douglas-fir fertilized with N (15),
Scots pine fertilized with macro- and micronutrients (42), and
loblolly pine fertilized with N and P (84). In the loblolly
pine study, regressions of growing season predawn xylem potential, measured at several dates, on soil water content
differed significantly among treatments, with fertilized trees
exhibiting less predawn stress than control trees at equal
soil water contents (Figure 1). Reduced mid-day water stress
(greater leaf water potential) was observed in fertilized

plots as compared to controls in both the Douglas-fir and Scots pine studies.

Several mechanisms may account for reduced tree water stress, despite reduced soil water content or potential, following fertilization. Both Brix (15) and Troth (84) cited the possibility of increased root growth associated with fertilization and Brix suggested less suberization of new roots under fertilized conditions; however, neither study included measurements of root growth. Any increase in rooting density at lower soil depths may have been especially beneficial to

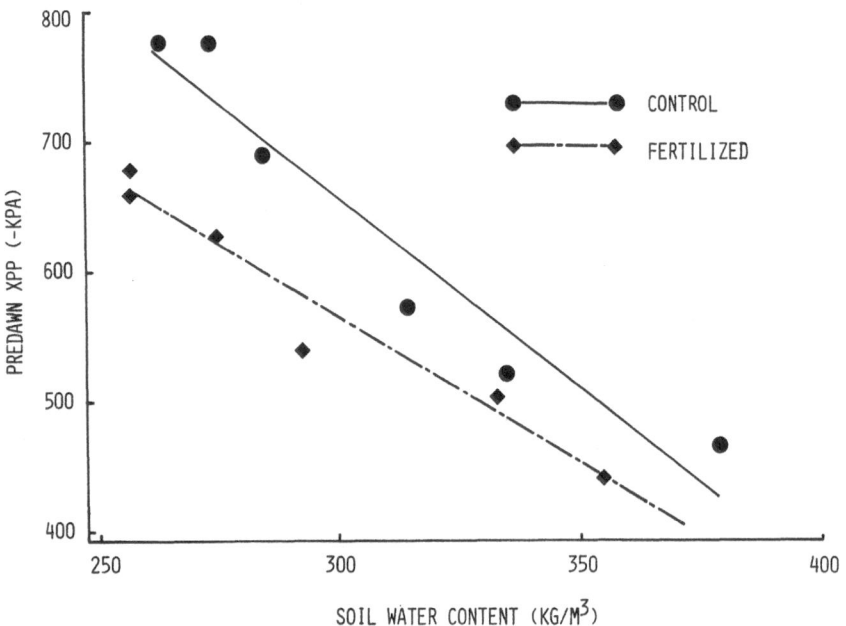

FIGURE 1. Effect of N + P fertilization on the relationship between soil water content and predawn xylem pressure potential in a 7-year-old loblolly pine plantation in Mississippi. Values are for 6 measurement dates during the first 2 summers after initiation of fertilization (from Troth 1983).

water uptake, since soil water contents increased with depth beneath both control and fertilized stands in the loblolly pine study (84). Reduced resistance to water flow between soil and leaves of fertilized trees, resulting from changes in either the permeability or cross-sectional area of the conducting system associated with increased radial growth, has also been hypothesized as a mechanism for altered tree-soil water relations. (42). Because stem sapwood is a major reservoir for water storage in trees (89, 90) additional sapwood in fertilized trees may also serve as an increased reservoir for water storage which may contribute to flow during periods of high evaporative demand (42). Differences in stomatal control or stomatal density among treatments could also contribute to differing tree-soil water relations.

Brix (15) found that net photosynthesis rates of both N fertilized and unfertilized Douglas-fir responded similarly to water stress. At equal levels of stress, shoots of fertilized trees displayed greater photosynthesis rates per unit of leaf area than shoots of control trees.

Observations of reduced soil water availability or content beneath fertilized plots imply that fertilization increases soil water use. This seems likely given the substantial increases observed in leaf production (16, 54, 58, 63). Bengtson and Voigt (11) noted that fertilization with N, P, and K increased total water use of slash pine seedlings by 30 to 80%, although fertilization also reduced water use per gram of dry matter produced by 25 to 50%. Reduced soil water content beneath fertilized plots may also result from increased interception and evaporation of precipitation associated with greater leaf area. Jarvis (44) emphasized that interception and evaporation are major components of site water balance in temperate coniferous forests. In the southern U. S., these components may be especially important during the

summer when rainfall is frequently of short duration and
temperatures are relatively high.

4.4 RESPONSE DURATION

On average, annual volume response of loblolly pine
stands to a single N fertilization peaks within 2 or 3 years
after application and declines to relatively low or undetect-
able levels within 6 to 10 years (Figure 2). Similar response
patterns following N fertilization have been reported for a
variety of species and ecosystems (61, 93). The duration of
response to P fertilizers may be substantially longer than
that to N fertilizers, particularly on strongly P deficient
sites in the southern Lower Coastal Plain (Figure 2). It
should be recognized, however, that duration of fertilizer

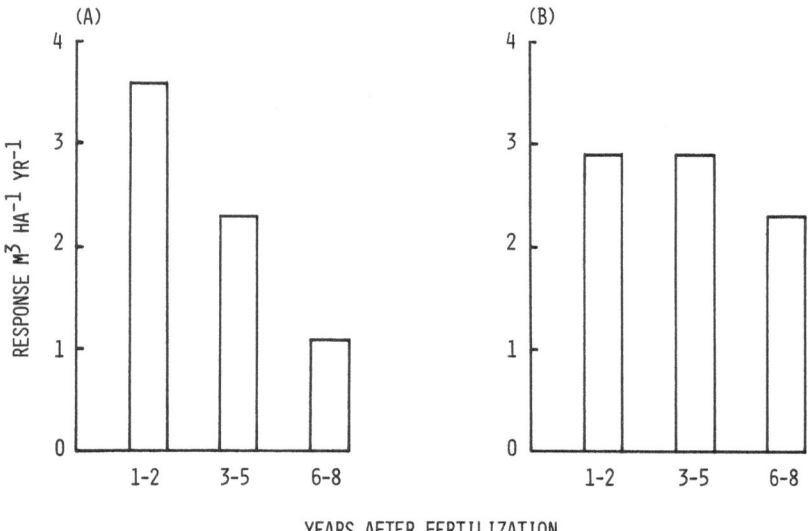

FIGURE 2. Duration of loblolly pine response to N fertilizer
in 24 stands (A) and P fertilizer in 14 stands (B) (Adapted
from North Carolina State Forest Fertilization Cooperative
Report No. 10) (70)

response varies greatly among individual stands. Fisher and Garbett (31) noted no decline in the response of some slash pine stands to N, P, or N+P fertilizers between years 5 and 8 after treatment, with increases in annual response occurring during this period in some situations.

Studies of soils beneath both loblolly pine (45) and Douglas-fir (41) stands indicate that increases in availability of soil N following urea applications may be of relatively short-term duration. After sharp initial increases, soil ammonium-N contents in fertilized plots declined rapidly and, after one growing season, had returned to levels similar to those in unfertilized plots. Soil nitrate-N increased after fertilization, but levels generally remained very low, although Heilman et al. (41) described one stand in which nitrate-N levels remained relatively high 24 weeks after N fertilization. These observations indicate that opportunities for greatly increased N uptake following fertilization are limited to the first growing season after treatment.

The duration of increased foliar N concentrations reflects the relatively short period of increased soil N availability. Nitrogen additions increased foliar N concentrations of both loblolly pine (66, 98) and Douglas-fir (16) sharply in the first season after application; however, these increases were no longer apparent by the end of the fourth season after treatment. Analysis of needles collected at monthly intervals following urea fertilization of radiata pine in New Zealand indicated that N concentrations in newly formed needles were increased only during the first growing season after treatment (48).

Yearly needle production of 24-year-old Douglas-fir peaked 2 to 3 years after N fertilization (Figure 3). Among the components of needle production, number of needles per shoot peaked in years 2 and 3, whereas number of shoots initiated peaked in years 3 and 4 (16). As expected, neither

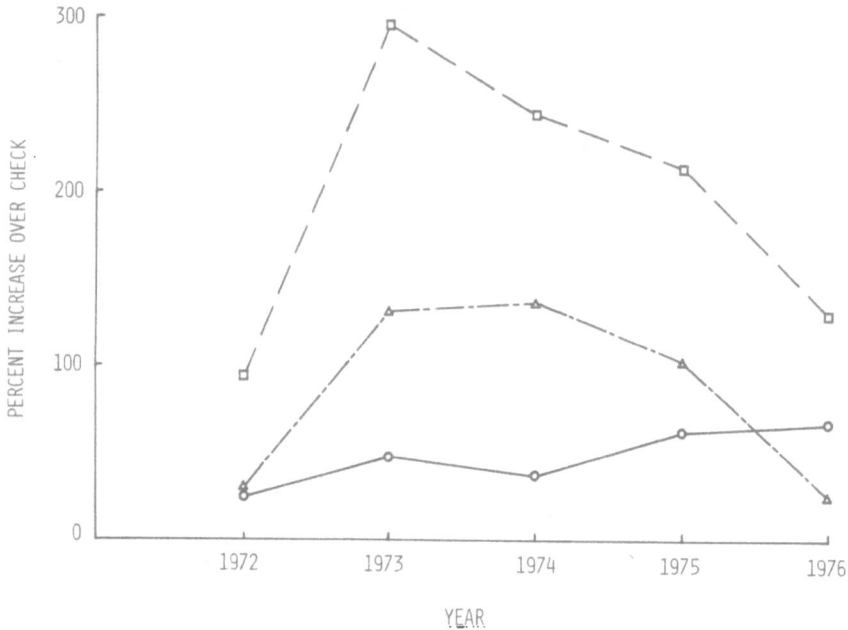

FIGURE 3. Percent increase in yearly needle production of Douglas-fir following application of 448 kg N/ha (△), thinning (o), and thinning + N fertilizer (□) treatments in the spring of 1972 (adapted from Brix 1981a)

component was increased in the first year for this species. Number of 1-year-old needles produced by radiata pine was increased by 80% 2 years after N fertilization, but by only 6% after the fourth year (58).

Brix (14) found that N fertilization of Douglas-fir increased net photosynthesis per unit of leaf area from early July of the first season to the end of July of the second season in shoots formed during the first season after treatment, but not thereafter. New shoots that expanded one year after fertilization exhibited increased rates of photosynthesis only during June. In a similar Douglas-fir stand, annual net aboveground biomass production per unit of foliage peaked in the first and second years following N applications to thinned and unthinned plots, respectively (18).

In recognition of the relatively short-term responses often observed in fertilizer trials, Miller (61) proposed that fertilizers generally benefit the trees, but do not lead to long-term improvements in site quality unless the amount of fertilizer retained on the site is large relative to the original nutrient capital. It was proposed that long-term gains in site productivity would result from applying N at very high rates or from P applied to very P deficient soils. As noted earlier P responses on P deficient sites are of greater duration than responses to N.

4.5 RESPONSE PREDICTION

Where growth responses to fertilizers vary widely among stands, an efficient, operational forest fertilization program requires identification and calibration of diagnostic variables to estimate the nutritional status of individual stands and their likely responses to fertilizer additions. Both stand and site measurements, as well as a wide array of tissue and soil analyses, have been tested for this purpose.

4.5.1 Stand and site variables

Volume response of unthinned loblolly pine plantations to N and N+P additions has been significantly related to stand basal area at the time of fertilization (6, 30). Where data were available from a wide range of stand and site conditions, initial basal area was better related to N response than any of a variety of nutritional variables, including soil and foliar N levels (6). Average volume response increased rapidly as treated basal area increased, with maximum response predicted at moderately high stand basal areas. Further increases in basal area resulted in reduced net volume response due to accelerated mortality in dense stands. Ballard (8) reported that greatest response of Douglas-fir to N also occurred at intermediate stocking levels and attributed reduced response to excessive competition at greater stocking levels and to incomplete site exploitation at lower stocking levels.

The relationship between initial basal area and volume response is apparently statistically significant because basal area is, in turn, correlated with variables such as stand leaf area and stand potential for production of additional leaf area, which play a major role in determining volume response (see Section 4.3.1). Basal area alone, however, is not a powerful predictor of fertilizer response, since this variable provides no measure of stand nutritional status. Ballard and Lea (6) found that basal area alone could account for only 28% of the variation among 31 locations in response of loblolly pine plantations to nitrogen.

Allen and Duzan (2) observed that the relationship between loblolly pine volume response and site index, an exhibited measure of site quality, may be either positive or negative. In situations where tree growth is limited primarily by nutrient availability, fertilizer response is apt to be negatively correlated with site index. Conversely, among sites where tree growth is most limited by soil water availability, response to fertilizers may be positively related to site index; that is, greatest response to nutrient additions can occur where water is least limiting. Ballard (8) suggested that a more detailed site quality function reflecting site water and nutrient regimes may prove more useful than conventional exhibited site index in predicting fertilizer response.

Fisher and Garbett (31) proposed a system of grouping soils for use in predicting the response of slash and loblolly pines to fertilizers. Their system, based on soil drainage class and depth to and nature of the subsoil, tended to group soils with similar soil water regimes. Although average response to fertilizers differed among groups, response also varied greatly within groups (49).

General relationships between fertilizer response and variables such as basal area or site index and use of soils groupings can guide managers of large industrial holdings in the selection of stands which are likely to respond to fertilization at some average level. However, response can be expected to vary widely and some non-responding or weakly responding stands will be fertilized. More reliable prediction of single stand response will require diagnostic procedures to estimate the extent to which nutrition limits stand productivity.

4.5.2 Foliar analysis

Although a wide variety of tissues and analyses have been tested as predictors of fertilizer response, total foliar N and P concentrations are widely used to quantify the nutritional status of forest trees. In conifers, these analyses are typically completed on fully formed needles collected from the upper part of the crown in a full sun position during late summer, autumn or winter. The age of needles collected, sampling date, crown position, and a variety of other factors significantly affect both the analytical results obtained (25, 59, 97) and the magnitude of differences among sites (74). Year to year differences in foliar nutrient concentrations have been associated with differences in temperature and rainfall prior to sampling (12, 65).

Foliar analysis has been a useful technique for definition of very P deficient loblolly and slash pine stands on Lower Coastal Plain sites in the southeast. Wells and Crutchfield (95) and Wells et al. (96) demonstrated that P additions at planting time produced consistent, strong increases in early height growth of loblolly pine where foliar P concentrations in control trees were less than 0.10%. Average response was near zero where foliar P concentrations of control trees averaged 0.12 to 0.14%. In slash pine plantations, significant response to P, applied near planting time, occurred where foliar P concentrations were less than 0.085 to 0.095% (7).

Few successful attempts to quantitatively relate foliar N or P concentrations to fertilizer response in established stands have been reported. Critical foliar concentrations have been suggested for several species (1), but many of these values have not been rigorously calibrated and should be used with caution. Another disadvantage in the use of critical levels is that these values do not reflect the continuous relationship between tissue nutrient concentrations and tree nutritional status. In addition, needle concentrations may indicate little about the magnitude of fertilizer response unless other stand and site characteristics can also be interpreted. Ballard and Lea (6) developed a model for predicting response of loblolly pine plantations to N which utilized foliar N concentrations in conjunction with initial basal area and site index (Figure 4). The authors speculated that foliar N concentration provided a measure of the nitrogen status of the stand, basal area an indication of the stands capacity for added growth, and site index an indication of the extent to which other site factors, including soil moisture, were limiting. The model, however, accounted for less than 50 percent of the variation in N response among sites.

The ratio of foliar % N to foliar % P was greatly superior to use of foliar N alone in classifying N responsive slash pine stands by discriminant analysis based on 5-year response information (27). A critical N/P ratio of 14 to 15 was suggested for operational use. The trials utilized for

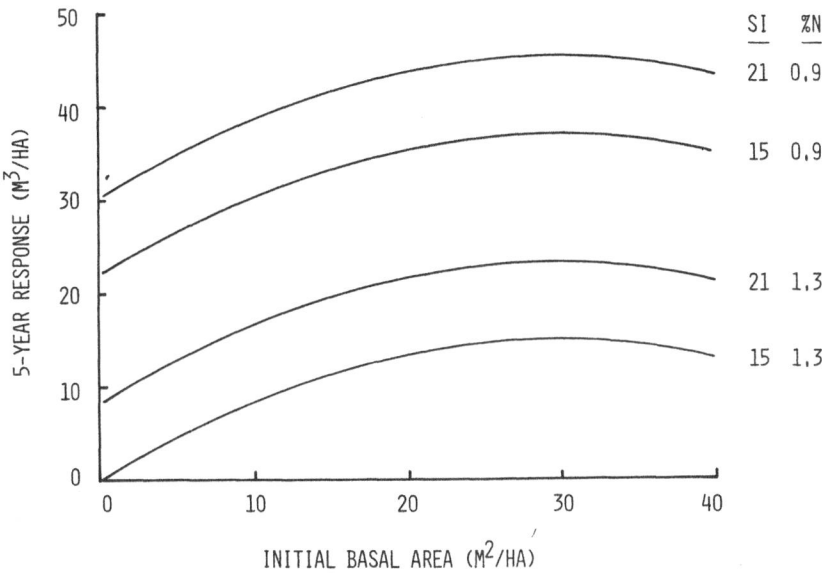

FIGURE 4. Relationship between predicted loblolly pine 5-year volume response to N and stand basal area, site index (m) and foliar % N (adapted from Ballard and Lea 1981)

this analysis were located in the southeastern Lower Coastal Plain where both N and P frequently limit growth. Consequently, a measure which incorporated information about both nutrients was superior to use of either alone in classifying stand response to N. Interestingly, foliar N/P ratios did not correctly classify the same stands as responsive or nonresponsive when 8-year response measurements were used, indicating that foliar N and P levels at time of fertilization provide better estimates of the short-term level of response, than of response duration.

Relationships between foliar N and sulfate-S concentrations have been explored for use in selecting N responsive
Douglas-fir stands (85, 86). Conceptually, relatively high
levels of sulfate-S in foliage were expected to indicate that
N availability was limiting tree growth, since utilization of
N and S are biochemically linked. This procedure resulted in
a correct classification of 17 of 19 Douglas-fir N trials as
responsive or nonresponsive using a critical foliar sulfate-S
concentration of 0.04% (86). Soluble S and P have also been
reported to accumulate in tissues of Norway spruce (Picea
abies (L.) Karst.)and Austrian black pine (Pinus nigra var.
nigra Arnold) when N was limiting protein synthesis (24).

Optimum foliar N concentrations apparently vary with tree
size or age (27,64). A series of experiments with Corsican
pine indicated that during the years prior to canopy closure,
optimum N concentration declined from 3.3 percent in very
young seedlings to 1.5 percent in forest trees 2.0 to 2.5 m in
height (64). At all growth stages the N concentrations
associated with optimum height growth were less than those
required for maximum diameter, volume, or weight increment.

Several researchers have attempted to strengthen the use
of foliar analysis for prediction of fertilizer response by
collecting foliage from both unfertilized and fertilized trees
at the end of the first growing season after treatment (83,
91, 92). First-year effects of fertilizers on foliar nutrient
concentrations, nutrient contents (mg/needle or fascicle), and
weight of needles or fascicles are then determined and graphically displayed. The technique is illustrated in Figure 5
using foliar data and 5-year volume response information from
a loblolly pine fertilization trial in the Ouachita Mountains
in Arkansas. Diagonal lines indicate increases in fascicle
weight. For the study illustrated, addition of N increased
foliar N concentration and content, but did not significantly
alter fascicle weight. Addition of N and P together increased

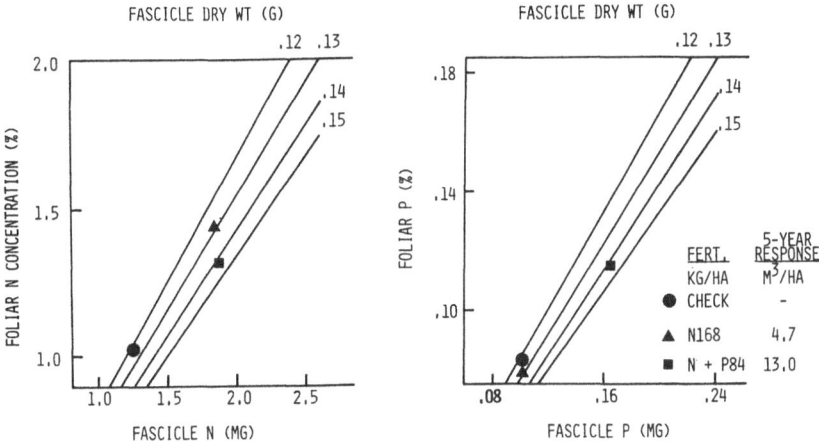

FIGURE 5. Effect of N and N+P fertilization on loblolly pine fascicle weight, foliar N and P concentrations, and fascicle N and P contents in September of the first season after application. Results are from a 5-year-old plantation in Arkansas.

fascicle weight and N and P concentrations and contents, and produced a much greater 5-year volume response than did N alone. Additional research is needed to determine whether a combination of the foliar responses illustrated can be used to predict fertilizer response more reliably than initial concentration data alone.

Concentrations of total soluble nitrogen or its components, such as arginine, have been more sensitive than total foliar N to N fertilization in various tissues of several coniferous species including Douglas-fir (88), loblolly pine (29), and slash pine (9). Multiple regressions using data from six Douglas-fir N fertilizer trials showed that basal area growth following treatment was correlated with initial stand variables, but that correlations could be improved by adding measures of soluble N compounds, particularly soluble N compounds in roots (88). A major difficulty in the use of soluble N and its components has been the high degree of variability associated with these measurements (29, 87). N source also affects free amino acid composition of tissues (87).

Analyses of fresh fallen needle litter have been of interest due to the difficulty of obtaining needles from upper crowns of older trees. Needle N concentrations in both Corsican pine (62) and black spruce (_Picea_ _mariana_ (Mill.) B.S.P.) (53) were not well correlated with concentrations in needle fall of the same year, but were closely correlated with levels in needle fall of the following season.

4.5.3 Soil analysis

Total soil N is not generally used as a measure of stand nutritional status or as an indicator of fertilizer response, since total N does not provide a measure of either N availability or the rate of N mineralization from organic matter. Total forest floor N beneath Douglas-fir has been negatively related to N response in thinned stands, but not in unthinned stands, suggesting that increased decomposition and N mineralization rates in thinned stands resulted in reduced fertilizer response (72).

In recent years, efforts to characterize soil N status have focused on evaluation of various procedures for assessing soil N availability, including measures of potentially mineralizable N. These procedures include both biological incubations, under either aerobic or anaerobic conditions, and a wide array of chemical soil tests (46). Ion exchange resin bags, buried in forest soils for several months prior to recovery, have also been used to evaluate soil nutrient availability (13, 40).

Shumway and Atkinson (80) found that soil N mineralized during an anaerobic incubation and percent stocking together accounted for 86% of the variation among 16 Douglas-fir stands in response to applied N. In ponderosa pine (Pinus ponderosa Laws.) stands, soil N mineralized during a standardized laboratory incubation was correlated with N mineralized in field incubations, site index, and foliar N concentrations (73). However, field soil temperatures had an important effect on the relationship between N mineralized in the laboratory and that mineralized in the field. Percent growth response of California stands to urea-N was also negatively related to mineralizable N, although relationships varied among frigid and mesic soils (74). In contrast, N fractions mineralized from surface soil by 2N KCl, anaerobic or aerobic incubation, autoclaving, and dilute acid were not correlated with 5-year volume response to N, site index, or foliar N levels in 31 southern pine plantations (50).

Extractable soil P has proven useful for identifying sites on which loblolly and slash pines are likely to respond to P applied at planting time (7, 96). Comerford and Fisher (26) found that surface soil extractable P (0-20 cm) did not adequately discriminate fertilizer responsive sites based on 12-year measurements, but that analysis of extractable P in samples collected to a total soil depth of 60 cm did provide acceptable discriminatory power. Ballard and Pritchett (7) also noted that the predictive ability associated with analysis of samples down to a soil layer which restricted rooting was greater than that associated with the analysis of surface soil samples. In 3- to 5-year-old stands, analysis of foliar samples from trees in check plots was superior to soil tests for prediction of P response (7, 96).

4.6 GENOTYPIC VARIATION

Within species variation has been described for a wide variety of tree responses to environmental gradients, including variation among progenies of select trees in their responses to varied nutrient regimes. A substantial proportion of all tree seedlings now planted on commercial forest lands are grown from seed collected from large production orchards. Where seed lots are kept separate by parent tree, all seedlings in a stand may have a common female parent. In future years, it is likely that increased use will be made of both controlled pollination, to produce seed where both parents are known, and, eventually, vegetative propagation. Fertility of forest sites on which these progenies are planted is generally low and many of the stands established will respond well to fertilizers. Consequently, one might anticipate that geneticists and nutritionists would have a strong interest in genotype x nutrition interactions. However, it appears that interest in these interactions has diminished and that relatively few forest researchers are now active in this area.

This lack of activity can be partially attributed to an attempt by tree breeders to select broadly adapted progenies which perform well across a wide range of sites, displaying low levels of genotype x environment interaction. Another important deterrent to operational use of selections, designed specifically for use on nutrient poor sites, is the increased cost and complexity of maintaining additional research, progeny testing, and breeding programs, and possibly separate seed orchards for genotypes with specialized nutritional responses. In cases where low levels of soil nutrient availability dictate improved nutrition to achieve acceptable yields, a solution through breeding requires not only that sufficient genetic variability in the desired attributes be present in the population, but that biological, ecological, or economic factors make a genetic solution more desirable than fertilizer additions. Several investigators have examined variability in growth among conifer genotypes in relation to nutrition. Goddard et al. (37) described height growth responses of 3-year-old open-pollinated slash pine families grown with and without fertilizers on poorly drained, acidic soils in north Florida. Some families grew poorly when un-fertilized, but responded strongly to fertilizers. Others which grew well without fertilizers displayed little response to treatment. Families which exhibited both superior un-treated growth and strong response to fertilizers were also cited. However, 8 to 10-year measurements of slash pine progenies at 4 locations revealed almost no family differences in response to fertilizers applied at planting, despite strong, general responses to fertilization (78). Control-pollinated families of loblolly pine have generally not dis-played significant family x fertilizer response interactions in field plantings (36, 37, 56), although fertilizer response estimates for families were most variable on wet, phosphate deficient sites where faster growing families tended to re-spond less than slow growing families (37). Variations among progenies in response to N have been demonstrated for both loblolly pine (77) and Douglas-fir (10), but only in short-term tests with very young trees. Examples are needed of response differences among progenies which will be economical-ly meaningful at harvest ages.

From a nutritional view point, the ideal genotype would be able to maintain rapid growth under low levels of soil availability, but also be able to respond strongly to in-creased availability (43). The potential mechanisms for these responses can be divided into two broad classes: 1) dif-ferences in root growth, geometry, or mycorrhizal infection which would improve the capability for nutrient uptake and 2) physiological differences which would result in improved up-take, distribution, or utilization of nutrients within the tree (36, 68). Graham (38) divided mechanisms by which nutrient efficiency can be achieved into five categories: 1) better root geometry, 2) faster specific rate of absorption at low ion concentrations, 3) chemical modifications within the rhizosphere to increase nutrient availability, 4) improved distribution within the plant and 5) superior utilization or a lower functional requirement for the nutrient.

Tree breeding to improve internal physiological mecha-nisms responsible for distribution and utilization of N and P would be seriously limited by a lack of knowledge regarding both the key mechanisms and the degree to which they vary among genotypes. Even in agronomic crops, the genetics of N and P efficiency are not well understood, largely because these elements participate in many plant processes and it is difficult to separate efficiency from growth attributes (38). Variations in physiological attributes of roots have been ob-served among genetic sources of conifers, although these were not examined in relation to nutritional responses (3, 57). It seems likely from such observations and agronomic experience that variability exists among tree genotypes in many of the physiological processes and enzyme systems which regulate nutrient uptake, distribution, and utilization.

Nambiar (68) argued that, at present, there is greater potential for manipulating the efficiency of tree nutrient uptake by selecting and breeding genotypes which exhibit superior root growth than by breeding for improved internal utilization. Nambiar and Cotterill (69) found pronounced dif-ferences among open-pollinated radiata pine families in the ability of seedlings to initiate and grow new roots. Cannell et al. (21) also reported family differences among loblolly pine seedlings in root to shoot relative growth rates and in relationships between total root length and shoot dry weight.

Families which produced the greatest 8-year volumes on a better-drained site, tended to have high root to shoot relative growth rates as seedlings, suggesting that superior volume producers avoided water stress by producing extensive root systems (21). If the energy cost of roots in such genotypes is not excessive, their use may confer improved growth potential through increased capacity for water and nutrient uptake. Since P and NH_4 ions are poorly mobile in soils, length of absorbing roots would be expected to be an important variable influencing their uptake.

Although mycorrhizal relationships have been extensively studied and their importance to uptake of nutrients, particularly P, is well established, differences among host families in their ability to form mycorrhizae and possible interactions among host and fungal genotypes have received little study. Marx and Bryan (55) reported that 1 of 4 slash pine families inoculated with Pisolithus tinctorius had significantly fewer mycorrhizal feeder roots than the other families. The ability of Thelephora terrestris to form ectomycorrhizal associations apparently was not influenced by host seedling genotype. Differences were also evident among host genotypes in the response of root numbers and foliar weights to fungal symbionts. The contribution of mycorrhizal associations to genotype x nutrition interactions in field studies with tree species is not known.

Agronomic researchers have examined mechanisms associated with P uptake efficiency in the belief that available P in many soils, particularly where P fertilizers have been applied, could support high growth rates of P efficient genotypes without further fertilization. Reviews by Clark (23) and Graham (38) cite a variety of plant responses associated with differential P uptake within agronomic species, including differences in rhizosphere pH, citrate release from roots, root length, maximum influx rates (Imax), Michaelis-Menton constants, and P concentrations remaining in solution when net influx was zero. Genotypes also differed in P distribution among shoots and roots and ability to retranslocate P from inactive to active tissues (23).

Goddard et al. (37) described a study conducted to gain some insight into the mechanisms by which some slash pine families were able to respond to P added to P deficient soils while others could not. Six families were selected, all of which displayed rapid growth without fertilization; 3 were responsive to added P and 3 were not. When soil P levels in a greenhouse trial were increased from 5 to 25 ppm, families responsive to P additions in field trials exhibited increased root growth. Families not responsive in the field displayed no root growth response. Experiments with radioactive P demonstrated that families responsive to P additions in the field also displayed a greater increase in P content of shoots when P availability was increased than did non-responsive families.

4.7 IMPLICATIONS FOR RESEARCH

In reviewing forest nutrition, it is apparent that much information has been gathered, but that important questions remain which seriously limit the forest manager in his ability to identify the level of nutrient stress in stands and to develop management plans which will minimize growth reductions due to suboptimal nutrition. Research in the following areas should be especially rewarding in terms of an improved ability to identify and manage nutrient stress:

- Further development is needed of procedures for diagnosis of the nutritional status of individual stands and estimation of response to fertilizer additions. It should be recognized that short- and long-term responses may be only loosely correlated and that variables which work well in predicting short-term response may not be suitable for estimation of

long-term effects (26, 27, 31). Factors regulating the duration of response are poorly understood. Procedures are also
needed to simultaneously evaluate the effects of both water
and nutrients on site productivity so that managers can assess
the probable effect of site water regime on response to improved nutrition. Variation in water availability likely
accounts for a significant portion of the observed variation
in fertilizer response.

   - Studies are needed of the effects of nutrition on tree-
soil water relations. Available evidence indicates that optimum nutrition can reduce tree water stress (15, 42, 84). The
interacting effects of fertilization and practices such as
thinning and competition control, which reallocate water use,
on tree-soil water relations should also be examined to provide guidelines for the role of fertilization in managing both
nutrient and water stress under varied stand and site conditions.

   - Better definition is needed of the relationships between internal nutrient availability and levels of key growth
processes. Brix (17) has related photosynthesis rates to
foliar N concentrations, but little is known of the relationships between internal N availability and numbers of second
order branches, numbers of leaf primordia initiated, or needle
size which are important components of crown response. The
effects of nutrition on dry matter partitioning among differing tree components, especially above- vs. belowground are
also uncertain.

   - Attempts to identify genotypes which differ in root
growth characteristics and to relate these differences to responses to both soil nutrients and water will be of great
interest. Uncertainties which surround the energy costs associated with rapid root growth rates, fine root turnover, and
root maintenance should also be considered.

REFERENCES

1. Allen, H.L., and T.M. Ballard. 1986. Fertilization: a silvicultural tool to enhance forest stand growth and value. J. Forestry 84:(In press).
2. Allen, H. L., and H. W. Duzan. 1983. Nutritional management of loblolly pine stands: a status report of the North Carolina State Forest Fertilization Cooperative. Pages 379-384. In: IUFRO Symposium on Forest Site and Continuous Productivity (R. Ballard and S. P. Gessel, eds.). USDA For. Serv. Gen. Tech. Rep. PNW-163. 406 p.
3. Allen, R. M. 1969. Racial variation in physiological characteristics of shortleaf pine roots. Silvae Genet. 18:40-43.
4. Axelsson, B. 1983. Methods for maintenance and improvement of forest productivity in northwestern Europe. Pages 305-311. In: IUFRO Symposium on Forest Site and Continuous Productivity (R. Ballard and S. P. Gessel, eds.). USDA For. Serv. Gen. Tech. Rep. PNW-163. 406 p.
5. Baker, J. B., G. L. Switzer, and L. E. Nelson. 1974. Biomass production and nitrogen recovery after fertilization of young loblolly pines. Soil Sci. Soc. Am. Proc. 38:958-961.
6. Ballard, R., and R. Lea. 1981. Foliar analysis for predicting quantitative fertilizer response: the importance of stand and site variables to the interpretation. In: Proceedings XVII IUFRO World Congress. Kyoto, Japan.
7. Ballard, R., and W. L. Pritchett. 1975. Soil testing as a guide to phosphorus fertilization of young pine plantations in the Coastal Plain. Univ. Fla. Agric. Exp. Stn. Bull. 778 (Tech.), Gainesville. 22 p.
8. Ballard, T. M. 1984. A simple model for predicting stand growth response to fertilizer application. Can. J. For. Res. 14:661-665.
9. Barnes, R. L., and G. W. Bengtson. 1968. Some aspects of nitrogen nutrition and metabolism in relation to fertilizer responses in southern pines. Pages 58-63. In: Forest Fertilization Theory and Practice. Tennessee Valley Authority, Muscle Shoals, Alabama. 306 p.
10. Bell, H. E., R. F. Stettler, and R. W. Stonecypher. 1979. Family x fertilizer interactions in one-year-old Douglas-fir. Silvae Genet. 28:1-5.
11. Bengtson, G. W., and G. K. Voigt. 1962. A greenhouse study of relations between nutrient movement and conversion in a sandy soil and the nutrition of slash pine seedlings. Soil Sci. Soc. Am. Proc. 26:609-612.
12. Bickelhaupt, D. H., R. Lea, D. D. Tarbet, and A. L. Leaf. 1979. Seasonal weather regimes influence interpretation of Pinus resinosa foliar analysis. Soil Sci. Soc. Am. J. 43:417-420.

13. Binkley, D., and P. Matson. 1983. Ion exchange resin bag method for assessing forest nitrogen availability. Soil Sci. Soc. Am. J. 47:1050-1052.

14. Brix, H. 1971. Effects of nitrogen fertilization on photosynthesis and respiration in Douglas-fir. For. Sci. 17:407-414.

15. Brix, H. 1972. Nitrogen fertilization and water effects on photosynthesis and earlywood-latewood production in Douglas-fir. Can. J. For. Res. 2:467-478.

16. Brix, H. 1981a. Effects of thinning and nitrogen fertilization on branch and foliage production in Douglas-fir. Can. J. For. Res. 11:502-511.

17. Brix, H. 1981b. Effects of nitrogen fertilizer source and application rates on foliar nitrogen concentration, photosynthesis, and growth of Douglas-fir. Can. J. For. Res. 11:775-780.

18. Brix, H. 1983. Effects of thinning and nitrogen fertilization on growth of Douglas-fir: relative contribution of foliage quantity and efficiency. Can. J. For. Res. 13:167-175.

19. Brix, H., and L. F. Ebell. 1969. Effects of nitrogen fertilization on growth, leaf area, and photosynthesis rate in Douglas-fir. For. Sci. 15:189-196.

20. Campbell, R. G., and J. L. Troth. 1981. Loblolly pine growth response to various sources of preplant phosphorus. Agron. Abstracts. Page 224.

21. Cannell, M. G. R., F. E. Bridgwater, and M. S. Greenwood. 1978. Seedling growth rates, water stress responses and root-shoot relationships related to eight-year volumes among families of _Pinus taeda_ L. Silvae Genet. 27:237-248.

22. Carlson, W. C., and C. L. Preisig. 1981. Effects of controlled-release fertilizers on the shoot and root development of Douglas-fir seedlings. Can. J. For. Res. 11:230-242.

23. Clark, R. B. 1983. Plant genotype differences in the uptake, translocation, accumulation, and use of mineral elements required for plant growth. Plant and Soil 72:175-196.

24. Clement, A., and S. P. Gessel. 1985. N, S, P status and protein synthesis in the foliage of Norway spruce (_Picea abies_ (L) Karst) and Austrian black pine (_Pinus nigra_ Arnold var. _nigra_). Plant and Soil 85:345-359.

25. Comerford, N. B. 1981. Distributional gradients and variability of macroelement concentrations in the crowns of plantation grown _Pinus resinosa_ (Ait.). Plant and Soil 63:345-353.

26. Comerford, N. B., and R. F. Fisher. 1982. Use of discriminant analysis for classification of fertilizer responsive sites. Soil Sci. Soc. Am. J. 46:1093-1096.

27. Comerford, N. B., and R. F. Fisher. 1984. Using foliar analysis to classify nitrogen-deficient sites. Soil Sci. Soc. Am. J. 48:910-913.

28. Comerford, N. B., R. F. Fisher, and W. L. Pritchett. 1983. Advances in forest fertilization on the south-eastern Coastal Plain. Pages 370-378. In: IUFRO Symposium on Forest Site and Continuous Productivity (R. Ballard and S. P. Gessel, eds.). USDA For. Serv. Gen. Tech. Rep. PNW-163. 406 p.

29. Cotrufo, C., and C. G. Wells. 1984. Some possible tissue assay methods for N nutrition assessment of Pinus taeda L. Commun. in Soil Sci. and Plant Anal. 15:1391-1407.

30. Duzan, H. W., H. L. Allen, and R. Ballard. 1982. Predicting fertilizer response in established loblolly pine plantations with basal area and site index. Southern J. Appl. For. 6:15-19.

31. Fisher, R. F., and W. S. Garbett. 1980. Response of semi-mature slash and loblolly pine plantations to fertilization with nitrogen and phosphorus. Soil Sci. Soc. Am. J. 44:850-854.

32. Gent, J. A., Jr., H. L. Allen, R. G. Campbell, and C. G. Wells. 1986a. Magnitude, duration and economic analysis of loblolly pine growth response following bedding and phosphorus fertilization. Southern J. Appl. For. 10:(In Press).

33. Gent, J.A., Jr., H.L. Allen, and R.G. Campbell. 1986b. Phosphorus and nitrogen plus phosphorous fertiliation in loblolly pine stands at establishment. Southern J. Appl. For. 10:(In press).

34. Gessel, S. P., E. C. Steinbrenner, and R. E. Miller. 1979. Response of northwest forests to elements other that nitrogen. Pages 140-149. In: Proceedings, Forest Fertilization Conference (S. P. Gessel, R. M. Kenady, and W. A. Atkinson, eds.). Univ. Washington Inst. For. Res., College For. Res. Contribution No. 40, Seattle. 275 p.

35. Gill, R., and D. P. Lavender. 1983. Urea fertilization effects on primary root mortality and mycorrhizal development of young-growth western hemlock. For. Sci. 29:751-760.

36. Goddard, R. E., and C. A. Hollis. 1984. The genetic basis of forest tree nutrition. Pages 237-258. In: Nutrition of Plantation Forests (G. D. Bowen and E. K. S. Nambiar, eds.). Academic Press, London.

37. Goddard, R. E., B. J. Zobel, and C. A. Hollis. 1976. Responses of Pinus taeda and Pinus elliottii to varied nutrition. Pages 449-462. In: Tree Physiology and Yield Improvement (M. G. R. Cannell and F. T. Last, eds.). Academic Press, London.

38. Graham, R. D. 1984. Breeding for nutritional characteristics in cereals. In: Advances in Plant Nutrition (P. B. Tinker and A. Lauchli, eds.). 1:57-102.

39. Grier, C. C., K. A. Vogt, M. R. Keyes, and R. L. Edmonds. 1981. Biomass distribution and above- and below-ground production in young and mature Abies amabilis zone ecosystems of the Washington Cascades. Can. J. For. Res. 11:155-167.

40. Hart, S. C., and D. Binkley. 1985. Correlations among indices of forest soil nutrient availability in ferti-lized and unfertilized loblolly pine plantations. Plant and Soil 85:11-21.

41. Heilman, P. E., T. Dao, H. H. Cheng, S. R. Webster, and S. S. Harper. 1982. Comparison of fall and spring applications of 15N-labeled urea to Douglas-fir: I. Growth response and nitrogen levels in foliage and soil. Soil Sci. Soc. Am. J. 46:1293-1299.

42. Hillerdal-Hagstromer, K., E. Mattson-Djos, and J. Hellk-vist. 1982. Field studies of water relations and photo-synthesis in Scots pine. II. Influence of irrigation and fertilization on needle water potential of young pine trees. Physiol. Plant. 54:295-301.

43. Jahromi, S. T., R. E. Goddard, and W. H. Smith. 1976. Genotype x fertilizer interactions in slash pine: growth and nutrient relations. For. Sci. 22:211-219.

44. Jarvis, P. G. 1985. Increasing productivity and value of temperate coniferous forest by manipulating site water balance. Pages 39-74. In: Forest Potentials Producti-vity and Value (R. Ballard, managing ed.). Weyerhaeuser Company Science Symposium 4. Weyerhaeuser Co., Tacoma, WA. 301 p.

45. Johnson, D. W., N. T. Edwards, and D. E. Todd. 1980. Nitrogen mineralization, immobilization and nitrification following urea fertilization of a forest soil under field and laboratory conditions. Soil Sci. Soc. Am. J. 44:610-616.

46. Keeney, D. R. 1980. Prediction of soil nitrogen availa-bility in forest ecosystems: a literature review. For. Sci. 26:159-171.

47. Keyes, M. R., and C. C. Grier, 1981. Above- and below-ground net production in 40-year-old Douglas-fir stands on low and high productivity sites. Can. J. For. Res. 11:599-605.

48. Knight, P. J., H. Jacks, and R. E. Fitzgerald. 1983. Longevity of response in Pinus radiata foliar concentra-tions to nitrogen, phosphorus, and boron fertilisers. New Zealand J. For. Sci. 13:305-324.

49. Kushla, J. D., and R. F. Fisher. 1980. Predicting slash pine response to nitrogen and phosphorus fertilization. Soil Sci. Soc. Am. J. 44:1303-1306.

50. Lea, R., and R. Ballard. 1982. Predicting loblolly pine growth response from N fertilizer using soil-N availabi-lity indices. Soil Sci. Soc. Am. J. 46:1096-1099.

51. Ledig, F. T., and T. O. Perry. 1965. Physiological genetics of the shoot-root ratio. Pages 39-43. In: Forest Resource Decisions in a Changing Power Structure. Proc. Soc. Amer. For., Detroit. 234 p.

52. Linder, S., and B. Axelsson. 1982. Changes in carbon uptake and allocation patterns as a result of irrigation and fertilization in a young Pinus sylvestris stand. Pages 38-44. In: Carbon Uptake and Allocation in Subalpine Ecosystems as a Key to Management (R.H. Waring, ed.). Oregon State University Forest Research Laboratory, Corvallis, OR.

53. Mahendrappa, M. K., and P. O. Salonius. 1982. Nutrient dynamics and growth response in a fertilized black spruce stand. Soil Sci. Soc. Am. J. 46:127-133.

54. Maki, T. E. 1960. Some effects of fertilizers on loblolly pine. Pages 363-375. In: Trans. 7th International Cong. Soil Sci., Madison, WI.

55. Marx, D. H., and W. C. Bryan. 1971. Formation of ectomycorrhizae on half-sib progenies of slash pine in aseptic culture. For. Sci. 17:488-492.

56. Matziris, D. I., and B. J. Zobel. 1976. Effect.of fertilization on growth and quality characteristics of loblolly pine. Forest Ecology and Management 1:21-30.

57. McClurkin, D. C., J. T. McClurkin, and T. J. Culpepper. 1971. Cytochemical and tissue homogenate analysis of adenosine triphosphate in root tips of Texas "Lost Pines". For. Sci. 17:446-451.

58. Mead, D. J., D. Draper, and H. A. I. Madgwick. 1984. Dry matter production of a young stand of Pinus radiata: some effects of nitrogen fertiliser and thinning. New Zealand J. For. Sci. 14:97-108.

59. Mead, D. J., and W. L. Pritchett. 1974. Variation of N, P, K, Ca, Mg, Zn, and Al in slash pine foliage. Commun. Soil Sci. and Plant Anal. 5:291-301.

60. Menge, J. A., L. F. Grand, and L. W. Haines. 1977. The effect of fertilization on growth and mycorrhizae numbers in 11-year-old loblolly pine plantations. For. Sci. 23:37-44.

61. Miller, H. G. 1981. Forest fertilization: some guiding concepts. Forestry 54:153-167.

62. Miller, H. G., and J. D. Miller. 1976a. Analysis of needle fall as a means of assessing nitrogen status in pines. Forestry 49:57-61.

63. Miller, H. G., and J. D. Miller. 1976b. Effect of nitrogen supply on net primary productivity in corsican pine. J. of Appl. Ecol. 13:249-256.

64. Miller, H. G., J. D. Miller, and J. M. Cooper. 1981. Optimum foliar nitrogen concentration in pine and its change with stand age. Can. J. For. Res. 11:563-572.

65. Miller, W. F. 1966. Annual changes in foliar nitrogen, phosphorus and potassium levels of loblolly pine (Pinus taeda L.) with site, and weather factors. Plant and Soil 24:369-378.

66. Moehring, D. M. 1966. Diameter growth and foliar nitrogen in fertilized loblolly pines. USDA For. Serv. Res. Note SO-43. 3 p.

67. Nadelhoffer, K.J., J.D. Aber, and J.M. Melillo. 1985. Fine roots, net primary production, and soil nitrogen availability: a new hypothesis. Ecology 66:1377-1390.

68. Nambiar, E. K. S. 1985. Increasing forest productivity through genetic improvement of nutritional characteristics. Pages 191-215. In: Forest Potentials Productivity and Value (R. Ballard, managing ed.). Weyerhaeuser Company Science Symposium 4. Weyerhaeuser Co., Tacoma, WA. 301 p.

69. Nambiar, E. K. S., and P. P. Cotterill. 1982. Genetic differences in the root regeneration of radiata pine. J. Exp. Bot. 33:170-177.

70. North Carolina State Forest Fertilization Cooperative. 1981. Response trends over time following fertilization of loblolly pine stands: evaluating long-term gain from short-term response (RW#1, two-, five-, and eight-year results. NCSFFC Report No. 10. School of Forest Resources, North Carolina State University, Raleigh. 244 p.

71. Peterson, C. E., Jr., and S. P. Gessel. 1983. Forest fertilization in the Pacific northwest: results of the Regional Forest Nutrition Research Project. Pages 365-369. In: IUFRO Symposium on Forest Site and Continuous Productivity (R. Ballard and S. P. Gessel, eds.). USDA For. Serv. Gen. Tech. Rep. PNW-163. 406 p.

72. Peterson, C. E., P. J. Ryan, and S. P. Gessel. 1984. Response of northwest Douglas-fir stands to urea: correlations with forest soil properties. Soil Sci. Soc. Am. J. 48:162-169.

73. Powers, R. F. 1980. Mineralizable soil nitrogen as an index of nitrogen availability to forest trees. Soil Sci. Soc. Am. J. 44:1314-1320.

74. Powers, R. F. 1983. Forest fertilization research in California. Pages 388-397. In: IUFRO Symposium on Forest Site and Continuous Productivity (R. Ballard and S. P. Gessel, eds.). USDA For. Serv. Gen. Tech. Rep. PNW-163. 406 p.

75. Pritchett, W. L., and N. B. Comerford. 1982. Long-term response to phosphorus fertilization on selected southeastern Coastal Plain soils. Soil Sci. Soc. Am. J. 46:640-644.

76. Pritchett, W. L., and W. H. Smith. 1972. Fertilizer responses in young pine plantations. Soil Sci. Soc. Am. Proc. 36:660-663.

77. Roberds, J. H., G. Namkoong, and C. B. Davey. 1976. Family variation in growth response of loblolly pine to fertilizing with urea. For. Sci. 22:291-299.

78. Rockwood, D. L., C. L. Windsor, and J. F. Hodges. 1985. Response of slash pine progenies to fertilization. Southern J. Appl. For. 9:37-40.

79. Russell, R. S. 1977. Plant root systems: their function and interaction with the soil. McGraw-Hill Book Company (UK) Limited. London. 298 p.

80. Shumway, J., and W. A. Atkinson. 1978. Predicting nitrogen fertilizer response in unthinned stands of Douglas-fir. Commun. in Soil Sci. and Plant Anal. 9:529-539.

81. Squire, R. O., G. C. Marks, and F. G. Craig. 1978. Root development in a Pinus radiata D. Don plantation in relation to site index, fertilizing and soil bulk density. Aust. For. Res. 8:103-114.

82. Sutton, R. F. 1969. Form and development of conifer root systems. Commonwealth Forestry Bureau Technical Communication No. 7. Oxford, England. 131 p.

83. Timmer, V. R., and E. L. Stone. 1978. Comparative foliar analysis of young balsam fir fertilized with nitrogen, phosphorus, potassium and lime. Soil Sci. Soc. Am. J. 42:125-130.

84. Troth, P. S. 1983. The effect of annual fertilization of loblolly pine (Pinus taeda L.) on soil water depletion and tree water stress. Unpublished. Weyerhaeuser Company Technical Report No. 050-3202/1. 26 p.

85. Turner, J., M. J. Lambert, and S. P. Gessel. 1977. Use of foliage sulphate concentrations to predict response to urea application by Douglas-fir. Can. J. For. Res. 7:476-480.

86. Turner, J., M. J. Lambert, and S. P. Gessel. 1979. Sulfur requirements of nitrogen fertilized Douglas-fir. For. Sci. 25:461-467.

87. van den Driessche, R. 1974. Prediction of mineral nutrient status of trees by foliar analysis. Bot. Rev. 40:347-394.

88. van den Driessche, R., and J. E. Webber. 1977. Variation in total and soluble nitrogen concentrations in response to fertilization of Douglas-fir. For. Sci. 23:134-142.

89. Waring, R. H., and S. W. Running. 1978. Sapwood water storage: its contribution to transpiration and effect upon water conductance through the stems of old-growth Douglas-fir. Plant Cell Environ. 1:131-140.

90. Waring, R. H., D. Whitehead, and P. G. Jarvis. 1979. The contribution of stored water to transpiration in Scots pine. Plant Cell Environ. 2:309-317.

91. Weetman, G. F., and D. Algar. 1974. Jack pine nitrogen fertilization and nutritional studies: three year results. Can. J. For. Res. 4:381-389.

92. Weetman, G. F., and R. Fournier. 1982. Graphical diagnosis of lodgepole pine response to fertilization. Soil Sci. Soc. Am. J. 46:1280-1289.

93. Weetman, G. F., and R. M. Fournier. 1984. Ten-year growth results of nitrogen source and interprovincial experiments on jack pine. Can. J. For. Res. 14:424-430.

94. Wells, C. G. 1970. Nitrogen and potassium fertilization of loblolly pine on a South Carolina Piedmont soil. For. Sci. 16:172-176.

95. Wells, C. G., and D. M. Crutchfield. 1969. Foliar analysis for predicting loblolly pine response to phosphorus fertilization on wet sites. USDA For. Serv. Res. Note SE-128. 4 p.

96. Wells, C. G., D. M. Crutchfield, N. M. Berenyi, and C. B. Davey. 1973. Soil and foliar guidelines for phosphorus fertilization of loblolly pine. USDA For. Serv. Res. Pap. SE-110. 15 p.

97. Wells, C. G., and L. J. Metz. 1963. Variation in nutrient content of loblolly pine needles with season, age, soil, and position on the crown. Soil Sci. Soc. Amer. Proc. 27:90-93.

98. Zahner, R. 1959. Fertilizer trials with loblolly pine in southern Arkansas. J. For. 57:812-816.

5. FOREST PESTS: INFLUENCE OF FOREST MANAGEMENT PRACTICES ON PEST POPULATION DYNAMICS AND FOREST PRODUCTIVITY

T. EVAN NEBEKER, D. R. HOUSTON, AND J. D. HODGES

Professor, Department of Entomology, Mississippi State University, Mississippi State, MS 39762; Principal Plant Pathologist, Northeastern Forest Experiment Station, Hamden, CT 06514; and Professor, Department of Forestry, Mississippi State, MS 39762

ABSTRACT

Considerable interest now exists in understanding how the host tree, pest population, associated microorganisms, and the environment interact to determine if a pest outbreak will occur. However, most research in the past has focused on individual components which determine pest outbreaks. To address management of "pest" stress it will be essential that we develop a better understanding of the repertoire of host defenses, and how genetics and environment interact to control expression of these defense mechanisms. The objective of this chapter will be to (1) bring together information gained through several studies which have attempted to elucidate the interrelationships between host, pathogens, etc. and (2) illustrate how this information can be incorporated into a forest management program to lessen the stress on the residual forest and decrease the susceptibility to pest attack.

5.1 INTRODUCTION

Demands for the use of our forest resources continue to increase. To meet these demands and to protect this valuable resource, a better understanding of the interrelationships among forestry practices, pest populations, and management objectives will be required. Recently, we (13) have reviewed the literature regarding how thinning practices influence pest problems in southern pine. The management of pine forests in the southern United States, as well as other parts of the

country, has intensified as timber resource value has
increased and the need for sustained production has become
evident. In the rocky mountain coniferous forests, the
environmental consequences of timber harvesting have been
addressed (16). In the deciduous forests of the Northeast,
several pest related problems associated with past forestry
practices have been identified (4,6). In the latter case,
beech bark disease has been influenced by more subtle human
actions. American beech (Fagus grandifolia Ehrh.) traditionally has been held in low esteem by forest industries due
to its poor seasoning qualities. Discriminant harvesting of
the more merchantible hardwood associates such as sugar maple
(Acer saccharum Marsh.) and yellow birch (Betula
alleghaniensis Brit.) has resulted in the creation of forests
that are overly rich in large, mature, and overmature beech
trees. Such trees are especially susceptible to beech scale
(Cryptococcus fagisuga Lind.) infestations and to Nectria spp.
infection.

The conceptual framework, presented by Houston (5), for
dieback and decline of trees can be extended to the overall
intent of this chapter. Healthy trees are affected by
environmental stress (including environmental modification as
a result of forestry practices) and over time, trees altered
by that stress are invaded at some point by pest populations.
Hence, this chapter will deal with 1) a review of how cultural
treatments can be utilized to manage the level and/or duration
of pest populations in forest stands; 2) identification of the
opportunities that genetics (current and future) can provide
to reduce or increase tolerance of forest stands to pest
populations; 3) the possible blending and utilization of genetic and cultural treatments to reduce or tolerate pest populations; and 4) the future research needs and challenges in the
area of utilizing cultural treatments and genetics to better
manage both pest and environmental stress in forest stands.
We will draw heavily from southern and eastern forest examples
in both pine and hardwoods.

5.2 FOREST PEST MANAGEMENT GOALS

Overall, the objectives are to reduce pest populations and their effects (mortality, defect, growth loss, etc.) to levels below some previously established economic, sociopolitical, or esthetic criteria and to maintain them at acceptable levels or equilibria (12). Specific goals to accomplish these may be to 1) decrease the amplitude of the fluctuation (outbreaks) in the pest populations, 2) increase the interval between outbreaks, and 3) shorten the outbreak periods.

Development of long-term timber management goals must consider the potential impacts of insect and disease on productivity. Silvicultural and tree improvement programs designed to meet these goals must include steps to minimize pest related losses in productivity. Human and biological inertia insure that pests will be presented with relatively uniform hosts. Because of the inherent time-lag between discovery and implementation of new research results, we feel that research programs to gain information about the relative contribution of parentage versus environmental modifications (silvicultural practices) in maintaining pest-resilient plantings should be initiated as soon as possible.

5.3 CULTURAL TREATMENTS

A look at the history of forest stand development in the U.S. reveals that many of today's forests occur on abandoned agriculture land. In some of these, and in many others that were always forested, the selective harvesting that frequently occurred, left stands of poor quality trees often overly rich in one species or another. The frequency and severity of pest-related problems have increased in recent decades. Thus, the conditions favoring the buildup of both the beech scale that triggered beech bark disease in the Northeast and the insect defoliators that initiated a sugar maple decline called maple blight in the Lake States appear to have been created by selective harvesting (4,6). In the South, abandonment and subsequent planting of pines on millions of hectares of abandoned farmlands set the stage for outbreaks, in these now

overstocked stands, of the southern pine beetle (SPB),
Dendroctonus frontalis Zimm.

In pest management, one option always considered in
benefit/cost analyses is that of doing nothing (12). What are
the implications of doing nothing? From historical records,
one can expect levels of pest activity and host damage to
occur in proportion to resource availability.

5.3.1 Species Selection

The selection of species to be regenerated depends on
numerous considerations and a clear idea of the management
objectives. Resistant species and varieties are a possibility. In southern pines we know that loblolly (Pinus taeda
L.) and shortleaf (P. echinata Mill.) pines are more preferred hosts of the SPB than are slash (P. elliotti Engelm.)
and longleaf (P. palustris Mill.) pines (2). Further, within
a species we can elect to plant known resistant families or
"woods run" (genetic history unknown) trees. The continued
planting of loblolly pine insures a sustained resource for the
southern pine beetle.

Following stand opening in beech by a disease or salvage
operation, thickets of beech, often of root sprout origin,
may develop. These young stems are genetically susceptible,
as were their parents to the beech bark disease. But,
because young or small stems are not as physically suited for
heavy scale colonization, stem infestation and subsequent
Nectria infection is localized. Defects in the form of
discrete cankers develop and accumulate. In many long-
affected forests, trees occur that are free of beech scale
and defect. These trees, which have been shown to be
resistant (some are immune) to the beech scale, may occur as
isolated trees. Often, however, they occur in groups. Many
groups of resistant trees have been located and mapped. The
management objective here would be to 1) enhance or maintain
the levels of resistant beech trees; 2) reduce the levels of
susceptible trees; and 3) create a more diverse species composition (6). How diverse a forest is in species composition

and structure may determine both its susceptibility and its
vulnerability to diseases and insects (7). We still do not
know the genetic relationships of these resistant beech trees
or how they originated, which is something we must find out
before sound silvicultural prescriptions can be made.
Although it seems feasible to reduce losses to diseases and
insects by regulating forest diversity through silviculture
and species selection, few trials have been attempted, at
least in the eastern United States.

Terry, et al. (15) have focused on the idea of pine
species-genotype selection and presented the following
general guidelines for southern pine. Natural regeneration
should be used only when the desired species is growing on an
area at the desired stocking to provide favorable regeneration. Care should be taken to select the most vigorous pest
free individuals as seed trees for regeneration. The
following are examples where one species is preferred over
another:

- -Slash pine is severely damaged by ice storms and can
  become heavily infected with fusiform rust. Slash
  pine, therefore, should be planted within its natural
  range and on wet sites where it has a demonstrated
  superiority over loblolly pine.
- -Shortleaf pine is very resistant to fusiform rust, but
  in the seedling stage can be severely infested with tip
  moth (Rhyacionia spp.). Older stands may be susceptible
  to littleleaf disease (caused by Phytophthora cinnamomi
  Rands) on certain sites. Generally, loblolly will be
  preferred over shortleaf pine, except on very dry sites
  in the northern limits of loblolly's range where ice and
  cold damage is severe.
- -Longleaf pine is not susceptible to tip moth or rust,
  but can be seriously damaged in the seedling stage by
  brown spot needle blight (caused by Scirrhia acicola
  (Dearn.) Sigg.). It also remains in the "grass stage"
  for an extended time. However, because of excellent
  resistance to fire damage and fusiform rust, longleaf

may be preferred on drier sites within its range where
these problems are chronic.

-Longleaf pine and sand pine (P. clausa (Chapm.) Vasey)
are preferred on certain sandy, well-to excessively-
drained soils within their natural range.

Once the preferred species is selected the best seed
source needs to be selected. One must carefully evaluate the
risks associated with moving seed sources with potentially
faster growth rates to nonlocal areas, particularly to areas
where low-probability weather events might cause severe physiological stress (15). This may place the selected species
in a more vulnerable position.

5.3.2 Site Preparation and Early Stand Management

Once the species to be propagated has been selected, the
proper site preparation techniques and subsequent management
prescriptions must be chosen. Many options are available and
range from doing nothing to using intensive cultivation and
manipulation techniques. In general, we know that when
loblolly pine is selected and intensive site preparation
techniques are employed there is an increase in tip moth
problems (15, 1). Long term effects of site preparation and
early management on the stand and future wood products need
to be considered. Our understanding of this topic in relation to hardwoods is lacking in comparison to our knowledge
of pine.

5.3.3 Stand Modification

In general the best management goals in hardwood stands
should be to promote or maintain some species diversity, and
when modifying these stands, to consider the diversity of the
residual stock. The reduction in species diversity resulting
from highgrading of eastern hardwood stands has been implicated as an underlying cause of various pest problems.
Whenever management practices are likely to alter the stand
species composition, an evaluation of the effects on potential pest related problems should be made. In bottomland

hardwood stands of the South the relationship is not as well defined, but it appears that some diversity is desirable from the standpoint of timber production as well as pest resistance.

Literature about managing basal area with thinning is extensive for the major southern pine species. We have provided a treatise (13) on the concept of thinning, reviewed and summarized thinning research, surveyed current field practices, and related the positive and negative aspects of these practices to current or potential destructive agents.

Any thinning strategy must consider the potential hazards associated with such intensive silvicultural practices (14). Bark beetle infestations are often associated with poor tree vigor which, in turn, may be altered by thinning. Though vigor is difficult to quantify, radial growth rate can serve as a strong indicator of a tree's physiological condition. Factors that affect tree growth include age, genetic make up, stand density, soil texture and type, fertility, drainage patterns, and stand disturbances associated with cultural practices. Poor tree vigor (poor growth) is usually associated with overstocking. Poor stand vigor can be improved by thinning, especially if directed at removing the lower crown classes (14). Thinnings can eliminate less vigorous or weakened individuals which are the prime targets for bark beetle attack. Bark beetles may attack the residual dominant and co-dominant trees, however, the removal of suppressed trees lowers between-tree competition. This enables the dominant trees to increase growth rates, upgrade stand vigor, and thus increase resistance to insect and disease attack.

Properly executed thinning will stimulate radial growth, reduce evapotranspiration, and increase rain throughfall. Reduction in evapotranspiration slows down the exhaustion of groundwater supplies and permits continued growth. Preventing severe water stress results in lower monoterpene concentration and higher levels of resin acids which could be involved in making the stand less attractive to beetles (3). Pine stands subjected to frequent flooding also become

attractive to bark beetle attack. In such stands, thinning
alone will not correct the problem. Additional forestry
practices such as drainage to divert excess water may be
necessary (14).

The distribution of slash, piled or scattered, needs to
be considered when planning a thinning strategy. A number of
pest species use the slash in which to reproduce. We have
observed, on numerous occasions, that stems of residual trees
are invaded by insects emerging from the slash piled near
their base. By scattering slash uniformly throughout the
stand, the potential for insect attack can be lessened. An
additional benefit of this practice is that such a layer of
slash can help distribute the weight of the woods machinery,
hence reducing the damage to roots and lessening soil
compaction.

5.4 GENETIC OPPORTUNITIES

We suggest that there are tremendous opportunities to
reduce the effects of stress on southern pine stands of the
future by properly matching species (genotypes), sites, and
susceptibility characteristics. In hardwood stands, increased
levels of resistance or tolerance to stress factors also
appear realizable. This would be accomplished through the use
of timber stand improvement operations and regeneration strategies which maintain or enhance species diversity and select
for tolerant or resistant clones, varieties, or individuals
within a species.

Evidence for genetic control of susceptibility to stress
effects in hardwoods comes primarily from plantation trials
with a few species including aspen, hybrid poplars, and black
walnut. Marked differences in susceptibility to weakly patho-
genic stem and root fungi and ambrosia beetles have been noted
between (different) clones, families, and provenances of these
species. Today, intensive evaluations of pest interactions
are being made for a large number of native and hybrid clones
or families planted on a wide variety of sites.

Relatively little is known about the genetic control of stress effects in most other native hardwood species growing under natural conditions. Studies with beech bark disease have shown that resistance to the beech scale stress agent exists within the native American populations (8). Repeated observations of infested trees over many years, together with the results of challenge trials, indicate that a continuum of resistance exists within a stand. Some trees will be completely immune; some will allow establishment of only sparce populations, but most will be extremely susceptible. This spectrum of resistance and observed patterns of bark colonization by the scale suggest that both biochemical and anatomical defense mechanisms are involved (9).

Insect-free (immune) trees occur in low numbers (12-15/Ha) and are usually isolated but some do occur in groups (8). Studies are now underway to determine (1) the genetic make-up (basis of immunity) of such trees (e.g. clone, half-sib, other) and (2) the heritability patterns of resistance in half-sib progeny of resistant parents. This information is necessary before silvicultural strategies can be devised to maintain or enhance resistance levels. Studies of how resistance and susceptibility to beech scale is distributed within natural stands may help explain the patterns of other stress-initiated problems in other hardwood tree species.

The role that past forestry practices played in creating susceptible forests to the beech scale was mentioned earlier. Practices that intentionally or inadvertently tend to reduce diversity of species or stand structure seem destined to increase stand susceptibility to stress-triggered problems. This is especially clear for problems such as beech bark disease that are initiated by host-specific biotic stress agents. Silvicultural approaches to reduce the overabundance of beech have been shown feasible in the Adirondacks (10). These approaches were combined with the use of herbicides to eliminate all advance regeneration (predominantly beech) with a two-cut shelterwood system in which species other than beech were favored as seed sources. This regeneration system, modified to retain all scale-resistant beech trees present for sources of seed or of vegetative root sprouts, has been put into trial in Maine.

The influence of species diversity in reducing stand susceptibility to stress-agents with broad host ranges is less obvious. Yet, even such pests as gypsy moth and forest tent caterpillar have host preferences, and, at least for gypsy moths, other stand and tree structural features appear to increase stand susceptibility. Research is currently underway in Appalachian oak-hickory forests to ascertain if susceptibility to gypsy moth defoliation can be reduced through silvicultural alteration of species composition and structure.

Similar arguments as above can be made for many of our forest pests. We might conclude from the literature (2, 11) for example, the ideal southern pine, from a forest manager's point of view, is one that is vigorously growing with a straight bole and a medium-sized crown. It would exude a lot of resin when attacked by a bark beetle. The resin would be relatively viscous and would flow for an extended period of time. Chemically, the resin would contain low monoterpene, high resin acid levels. Alpha-pinene content would be low. This tree would contain an adequate level of defensive chemicals ranging from toxic, low-molecular-weight compounds, to high-molecular-weight digestibility-reducing compounds. It would produce more of the same chemicals when exposed to insect attack. It is these types of traits that one should be considering in the tree improvement programs with a vision of preventing losses to pest species such as the bark beetles that are problems in older stands.

5.5 FUTURE RESEARCH NEEDS

An understanding of the mechanisms controlling resistance to pest attack is critical. It is essential that we understand the repertoire of the host defenses, the magnitude and expression of which are contingent on the genotypic and

phenotypic "vigor" of the individuals in the host population
(11). Further, it is essential that this information be provided to silviculturists and plant breeders for improving the
resistance within the host type. Key questions to be
addressed, such as what is the relative contributions of
parentage (genetic) versus environmental (climatic and silvicultural practices) in determining the magnitude of
resistance within a host type?, need to be answered.

The interactions between genetic and cultural treatments
are now just beginning to be addressed. A great deal of
emphasis needs to be placed on the elucidation of the host
defensive systems and to determine to what extent each system
is genetically controlled and whether they are environmen-
tally modifiable. The greatest challenge is the visualiza-
tion of the problem and the subsequent experimentation to
understand "stress" development in relation to the interaction
of the host tree, microorganisms, and pest populations.

REFERENCES

1. Hedden, R. L. and T. E. Nebeker. 1984. Integrated pest
   management in pine stands (0-5 years)--insects. In
   Branham, S. J. and G. D. Hertel (eds.). Proceedings
   Integrated Forest Pest Management Symposium, Athens,
   Georgia. pp. 39-53.
2. Hodges, J. D., W. W. Elam, W. F. Watson, and T. E.
   Nebeker. 1979. Oleoresin characteristics and susceptibility of four southern pines to southern pine beetle
   (Coleoptera: Scolytidae) attacks. Can. Ent. 111:
   889-896.
3. Hodges, J. D. and P. L. Lorio, Jr. 1975. Moisture
   stress and composition of xylem oleoresin in loblolly
   pine. For. Sci. 21: 283-290.
4. Houston, D. R. 1979. Spreading tree diseases: the hand
   of man. The Ecologist 4/5: 120-124.
5. Houston, D. R. 1981. Stress triggered tree diseases:
   the diebacks and declines. USDA For. Serv. NE-INF-41-81.
   36p.
6. Houston, D. R. 1981. Some dieback and decline diseases
   of northeastern forest trees: Forest management considerations. In Proc., The National Silviculture
   Workshop, Roanoke, VA, June 1-5, 1981. USDA For. Serv.,
   Div. Timber Mgmt., Washington, DC. pp. 248-265.
7. Houston, D. R. 1982. Diseases, insects, and forest
   diversity: silvicultural implications. Proc. Natural
   Diversity in Forest Ecosystems, Nov. 29-Dec. 1. Univ.
   Georgia, Athens, GA.

8. Houston, D. R. 1983a. American beech resistance to
    Cryptococcus fagisuga. In: Proc. IUFRO Beech Bark
    Disease Working Party Conference, Hamden, CT. Sept
    26-Oct. 8, 1982. USDA For. Serv. Gen. Tech. Rept. WO-37.
    pp. 38-42.
9. Houston, D. R. 1983b. Development in biological control
    of beech bark disease. In: Proc. 10th Intl. Congress of
    Plant Protection 1983. Brighton, England. Nov. 20-25,
    1983. Vol. 3: 1035-1041.
10. Kelty, M. J. and R. K. Nyland. 1981. Regenerating
    Adirondack northern hardwoods by shelterwood cutting and
    control of deer density. J. Forestry 79: 22-26.
11. Nebeker, T. E., C. A. Blanche, and J. D. DeAngelis.
    1984. Host/bark beetle/microorganism interactions. In
    Payne, T. L., R. F. Billings, R. N. Coulson and D. L.
    Kulhavey (eds.). History, status, and needs for entomo-
    logical research in southern forests. Texas A&M Misc.
    Pub. 1553, pp. 19-23.
12. Nebeker, T. E., R. F. Mizell, III, and N. J. Bedwell.
    1984. Management of bark beetle populations: Impact of
    manipulating predator cues and other control tactics. In
    Garner, W. J. and J. Harvey, Jr. (eds.) Chemical and
    Biological Controls in Forestry. ACS Symposium Series
    238. pp. 25-33.
13. Nebeker, T. E., J. D. Hodges, B. L. Karr, and D. M.
    Moehring. 1985. Thinning practices in southern pines-
    with pest management recommendations. USDA-Forest
    Service Tech. Bull. 1703. 36 pp.
14. Nebeker, T. E. 1985. Influence of forestry practices on
    bark beetle populations: a perspective. In Goyer, R. A.
    and J. P. Jones (eds). Insects and Diseases of Southern
    Forests. 34th Annual Forestry Symposium. Louisiana State
    University, Baton Rouge, LA. pp. 52-59
15. Terry, T. A., S. C. Cade, and J. H. Hughes. 1984.
    Integrated pest management during plantation establish-
    ment (0-5 years). In Branham, S. J. and G. D. Hertel
    (eds.) Proceedings Integrated Forest Pest Management
    Symposium, Athens, Georgia. pp. 97-114.
16. USDA Forest Service. 1980. Environmental consequences
    of timber harvesting in Rocky Mountain coniferous
    forests. USDA For. Serv. Gen. Tech. Rep. Int-90, 526p.
    Intermt. For. and Range Exp. Stn., Ogden, Utah 84401.

6. FOREST PESTS: THE ROLE OF PHLOEM OSMOTIC ADJUSTMENT IN THE DEFENSIVE RESPONSE OF CONIFERS TO BARK BEETLE ATTACK

P.J.H. SHARPE[1], R.J. NEWTON[2], AND R.D. SPENCE[1]
1 Professor and Research Assistant, respectively, Biosystems Research Group, Department of Industrial Engineering, Texas A&M University, College Station, Texas 77843
2 Associate Professor, Department of Forest Science and the Texas Agricultural Experiment Station, Texas A&M University, College Station, Texas 77843

ABSTRACT

Osmotic adjustment is a physiological process by which trees accumulate solutes in response to drought stress. It benefits the plant by maintaining turgor pressure and hence continued but reduced growth. Osmotic adjustment operates by the accumulation of soluble carbohydrates and amino acids within stressed tissues. An indirect consequence of osmotic adjustment is that it raises the nutritive value of affected tissues as a food source for herbivores. Osmotic adjustment without other metabolic changes therefore increases the vulnerability of stressed trees to pest attack.

Trees have evolved mechanisms for defense during periods of osmotic adjustment. It is the thesis of this chapter that osmotic adjustment provides a stimulus for the production of allelochemicals which defend the tree against herbivores. For example, conifer defense against bark beetles can be interpreted in terms of metabolic changes induced by osmotic adjustment. The physiological and anatomical changes associated with osmotic adjustment can be shown to provide an explanation for shifts between growth and differentiation phenomena identified in the early literature.

6.1 INTRODUCTION

It is generally accepted that a deviation from normal tree physiology results in increased vulnerability to pest and diseases (33). The degree of vulnerability of a tree to herbivore colonization is referred to as its susceptibility, which is the opposite of resistance. Susceptibility can be visualized as a spectrum; at one extreme trees are highly susceptible to attack, but at the other extreme they are highly resistant. Susceptibility, however, does not necessarily imply that herbivore attacks will occur. Other factors must be considered, such as the population dynamics of the herbivore, seasonal- and weather-related influences, and resource availability to the tree.

Because resources that can be used for growth and defense are finite, their optimum allocation by the tree is essential if the species is to survive and compete successfully with others. Townsend and Calow (63) suggest that there is a direct relationship between allocation of assimilates and tree vigor. Conditions that limit resource usage, such as drought, competition for light, and nutrient deficiency, alter the physiological status of the tree and ultimately its defensive strategy. Nutrients, energy, and substrates allocated to one function are generally unavailable for other processes. During periods of low stress, rapid growth is associated with low nutrient levels in the tissues (26). The tree is therefore less attractive to herbivores and can be less strongly defended.

Stress changes the nutritive content of the tissues, the attractiveness of the host tree, and the requirement for defensive chemicals. Figure 1 shows an influence diagram of these relationships. Stress causes two plant responses, osmotic adjustment leading to an increased nutritive value and the induction of allelochemical synthesis. Improved nutrient status of the tree facilitates pest outbreak by increasing pest vigor, whereas allochemical synthesis suppresses pest vigor. Under climatic stress, therefore, stand protection from pest outbreaks is maintained by a balance between enhanced nutrient content and allelochemical synthesis.

Evidence for the scheme outlined in Figure 1 has been presented in recent studies. Chew and Rodman (12) proposed that as a result of nutrient or drought stress, carbon that cannot be invested in growth is diverted to secondary

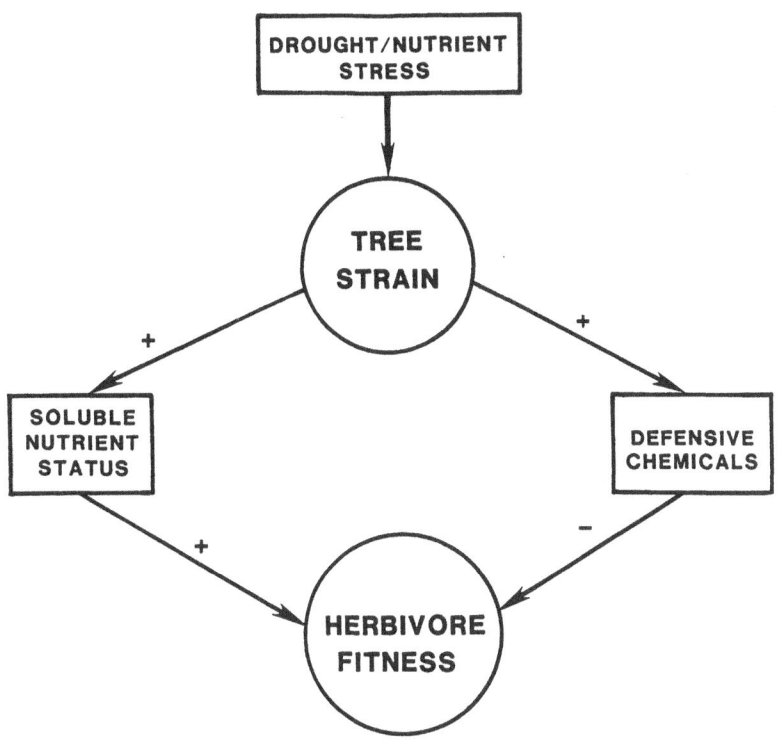

FIGURE 1. Influence diagram for tree-herbivore interaction (Strain concept follows that outlined by Levitt, 37). The plusses show that tree strain has a positive effect on nutrient status and the production of defensive chemicals, and that nutrient status benefits herbivore vigor. The minus shows that the production of defensive chemicals has a negative effect on herbivore vigor

metabolite (allelochemical) production. Bryant et al. (9) showed that observed increases in carbon-based (in contrast to nitrogen-based) allelochemicals occurred at times of the year and under environmental conditions during which there was a carbon surplus that could not otherwise be used for maintenance and growth. Nutrient fertilization diverted carbon to growth, however, and reduced the excess carbon available for the

116

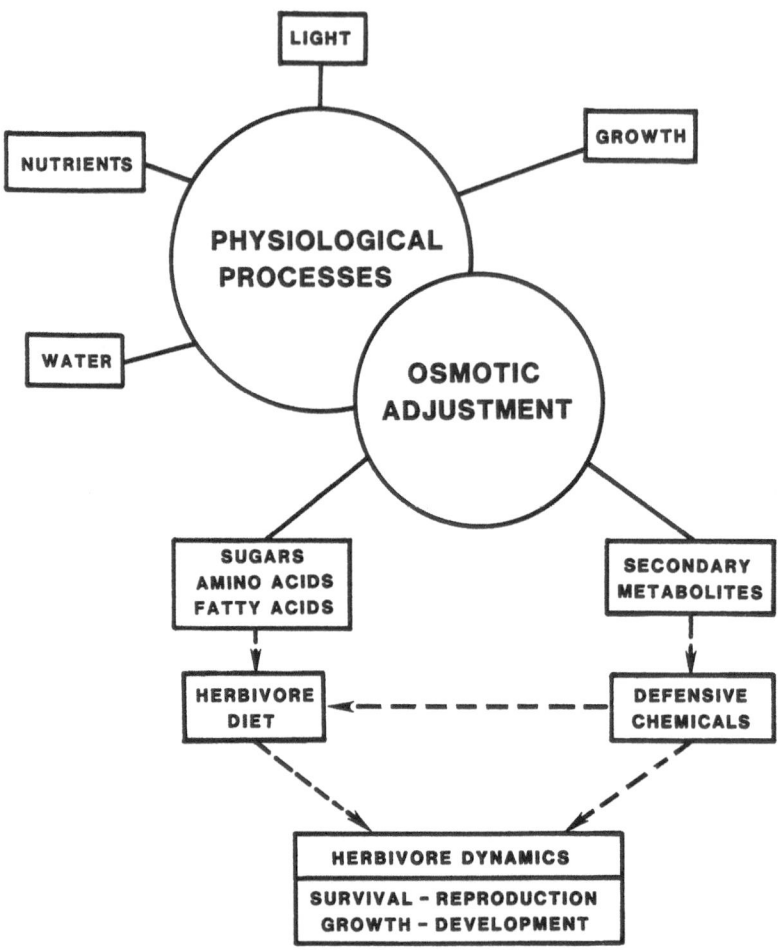

FIGURE 2. A conceptual diagram showing the key role of osmotic adjustment as a link between plant physiological processes and herbivore dynamics

production of allelochemicals. As a result of these observations, Tuomi et al. (64) hypothesized that changes in defensive allelochemicals result from the carbon/nutrient balance of the tree. In other words, changes in allelochemical concentrations are genetically-programmed physiological responses to changes in carbon/nutrient ratios, rather than direct active responses by the host tree to herbivory.

Neither Bryant et al. (9) nor Tuomi et al. (64) suggest a mechanism by which the production of allelochemicals is regulated. Osmotic adjustment (Fig. 2) has the appropriate characteristics to regulate allelochemical production. Osmotic adjustment may play two roles in stimulating production of allelochemicals: (i) as a strain-sensing mechanism and (ii) by increasing the availability of substrate. The following discussion suggests that osmotic adjustment is closely coupled to secondary metabolite production.

6.2 MECHANISMS OF OSMOTIC ADJUSTMENT

The term "osmotic adjustment" is used to describe solute accumulation in response to drought stress (44). Solutes that accumulate are sugars, amino acids, organic acids, and potassium and other inorganic ions such as chloride (30). These solutes have been shown to accumulate in leaves, roots and the inner bark (28).

The phloem translocation system modulates energy allocation within the tree and establishes the concentration of sink substrate for synthesis pathways in the stem. Extensive experimental and theoretical studies have shown that carbohydrate concentration increases in the phloem during drought stress (14, 15, 18, 19). Carbohydrate accumulation in the phloem is the first step in osmotic adjustment and has three effects: (i) It raises the cost of transporting materials because loading processes must operate against a higher sucrose or inorganic ion concentration. (ii) By increasing substrate concentration in the phloem, it raises the solute concentration in the epithelial resin duct cells of the phloem, thereby maintaining oleoresin pressure and favoring synthesis of oleoresins. (iii) It provides higher concentrations of carbohydrates and amino acids in the inner bark and xylem rays (26) for colonizing bark beetles and

associated fungi (55). Solute accumulation may have
counteracting effects on susceptibility to and suitability for
beetle infestation because the food value of the phloem is
increasing at the same time that synthesis of defensive
compounds is stimulated. Over short time periods, conditions
such as shading due to clouds passing across the sun that
reduce the degree of drought stress can also reduce oleoresin
exudation pressure (OEP) (65, 66). Longer-term changes in
drought stress, however, have not been correlated consistently
with OEP. Goeschl (20) proposed that the trade-off between the
increased food value of the phloem and solute accumulation in
the resin duct epithelial cells which maintain turgor in the
resin duct system was more important than any correlation that
may exist between OEP and defense against herbivores. Osmotic
adjustment therefore would break the relationship among degree
of drought stress, OEP field measurements, and herbivore
infestation success.

In addition to maintaining cell turgor, osmotic adjustment
allows the continued transport of photosynthate during drought
stress. The importance of osmotic adjustment for phloem
function was suggested by DeMichele et al. (14) and confirmed
by carbon-11 tracer studies (16, 19). Phloem translocation
dynamics during drought stress are complex, but their
relationship to solute adjustment and tree defense can be
summarized as follows: The metabolic energy cost for
translocation is low when solutes are transported at low
concentrations and high velocity. This condition cannot be
maintained under drought stress because negative pressures
develop first at the terminal end of the sieve tube (55).
Unlike xylem vessels and tracheids, the sieve tubes are
structurally delicate and plasmolyze or constrict as a result
of the balloon-like expansion of the surrounding parenchyma
cells, which can occur when extreme water deficit causes low
phloem pressures. Phloem transport is then blocked. The mean
pressure in the sieve tubes is based essentially on the
combination of the mean solute concentration and the water
potential of the nearest xylem tissue. Solute transport under

drought stress can be maintained functionally by two
mechanisms: (I) reduction in the phloem unloading rate in the
sink regions, especially in growing meristems and ray
parenchyma near resin ducts, and (II) increase in the phloem
loading rate. These two mechanisms are interdependent, with
mechanism I being sufficient for slight stress. As the drought
stress increases, however, both mechanisms I and II are invoked
for the following reasons:

    i) Negative pressures in the sieve tubes are overcome by a
       decrease in the phloem unloading rate, causing an
       increase in solute concentration in the phloem and a
       decrease in the mass flow velocity (20).

   ii) The resulting high concentration of solute inhibits
       loading and thus solute transport rate.

 iii) Active solute loading, particularly of sucrose, occurs
       only to the point at which high solute concentration
       causes excess sap viscosity and crystallization, which
       impede translocation (14).

   iv) Plants adapt to drought stress by establishing a
       relatively high concentration of solutes such as
       malate, amino acids, and inorganic ions in their cells.
       It is very likely that these osmotic solutes are
       recycled in the phloem-xylem system (4).

    Phloem-induced changes in carbohydrate and amino acid
concentrations throughout the plant account for (i) the
increased nutritive value of the tissue to the herbivore (11,
28, 51, 69, 70, 71), (ii) reduction in growth and promotion of
synthesis of allelochemical compounds, and (iii) a reduction in
the concentration of high molecular weight compounds that
reduce digestibility and an increase in the concentration of
low molecular weight, presumably toxic, carbon allelochemicals
(17, 28). The dynamics of these last two compounds have been
described in a model presented by Sharpe et al. (55).

6.3 OSMOTIC ADJUSTMENT STRATEGIES

    Osmotic adjustment has been demonstrated in Pinus taeda L.
(24, 54), Prosopis (46), Prunus, Ilex (31), Juglans nigra L.

(48), and Quercus (47). Osmotic adjustment increased with age
in the phreatophytic species, Prosopis, with mature shoots
having lower osmotic potentials than juvenile shoots (46). In
Prunus and Ilex, osmotic potentials of newly emerged leaves
were higher than those of mature leaves (31). Solute
accumulation during drought stress results in maintenance of
turgor and high leaf conductance in Quercus (47), mature leaves
of Prosopis (46), and several shrubs (25). Roots of Quercus
seedlings accumulated assimilate and adjusted osmotically more
effectively than did the leaves where the osmotic adjustment
was accompanied by increased root length.

The relative importance of increased food value and an
increase in chemical defense compounds to the vigor of the tree
depends upon tree genotype, growth history, and current
environmental conditions. Evolutionary selection pressure
works in two opposing directions. Selection for osmotic
adjustment allows increased growth and survival under drought
stress, but also results in an increase in susceptibility to
herbivore attack. Selection for synthesis of secondary
compounds reduces growth and decreases susceptibility. There
is evidence that different pine species have very different
strategies for balancing the demands for growth and defense.
According to Hodges et al. (29) and Blanche et al. (7), slash
pine (Pinus elliottii Englm.) appears to be better adapted to a
chemical defense against insects and diseases whereas loblolly
pine (Pinus taeda L.) appears to be better adapted to osmotic
adjustment and continued growth under drought stress
environments. These observations are consistent with loblolly
pine being the species best adapted to the western, drier
margins of the southern U.S., while slash pine is less drought
resistant but better adapted to the higher rainfall in the
central and eastern regions of the southern U.S.

6.4 TYPES AND DURATION OF STRESS

Osmotic adjustment has traditionally been related to one
type of stress only--drought stress. In this section we
discuss how other types of stress as well as the duration of

stress are likely to influence osmotic adjustment and synthesis of secondary compounds.

In the context of plant defense theory (51), stress is usually considered over a short time period and to be unpredictable in duration and intensity. Failure to distinguish between periodic short-term climatic stresses and long-term nutrient deficiencies has led to considerable confusion in understanding tree defense strategies. For example, tree genotypes that have evolved under long-term nutrient deficiencies have higher levels of polyphenols and organic acids than trees that have evolved under more abundant resource conditions (21, Cates, Horner, and Gosz, personal communication). Long-term resource deficiencies resulting from a lack of adequate water and/or nutrients appear to induce permanent states of osmotic adjustment and vulnerability to herbivore attack. High levels of carbon allelochemicals appear to be maintained as a deterrent.

Long-term reduction in light due to competition from neighboring trees has a very different effect on tree defense mechanisms than water and nutrient deficiencies. A decrease in light intensity or crown size reduces photosynthate production, reducing phloem carbohydrate concentration. It appears that synthesis of secondary defensive compounds is greatly reduced with a decrease in crown photosynthesis. Clements (13) states that the volume of oleoresin produced from southern pines is directly related to crown size. Waring and Pitman (68) relate leaf area and sapwood growth, both of which are directly related to sunlight interception by the host-tree, to host-tree vigor, and to herbivore resistance. Pines attacked by the southern pine beetle are generally associated with poor growth, smaller size, smaller crown, and thinner bark, relative to healthy trees (34, 35). In selecting trees susceptible to attack during field experimentation, morphological features such as yellowish needles, small cones, and sparse crowns were chosen (27). These features are all characteristics of sunlight deficiency.

Short-term stresses under natural conditions are limited to

extremes of water availability and temperature. Silvicultural
practices, however, can cause sudden increases in sunlight
availability (thinning) or nutrient availability (fertilization) (see Chapter 3, this volume). Little is currently known
about the effects of sudden changes in temperature, sunlight,
or nutrient availability on synthesis of secondary compounds
and resistance to herbivores. Experimental studies demonstrate
a range of responses to drought stress. Wadleigh et al. (67)
showed interactions among the degree of drought stress, growth,
and rubber synthesis in guayule. Growth decreased with
increasing drought stress, while rubber percentage yield
increased initially but then decreased as drought stress
increased. Black (6) suggested that under moderate drought
stress, photosynthate is diverted from growth to the formation
of secondary compounds such as rubber, essential oils, and
oleoresin constituents. These observations suggest that
osmotic adjustment resulting from moderate drought stress
stimulates synthesis of secondary compounds. With the onset of
severe stress conditions, osmotic adjustment and other
dependent processes are reduced. Futhermore, moderate drought
stress has a greater influence on osmotic adjustment processes
than on the supply of photosynthate.

6.5 GROWTH AND DIFFERENTIATION

The growth-differentiation concept was proposed by Loomis
(38) to complement the carbon-nitrogen balance theory and has
been invoked by Lorio and Hodges (43) to explain the seasonal
differences in synthesis and yield of oleoresin. Loomis (38)
divided plant development into three distinct phases: cell
division, cell enlargement, and cell differentiation. The
first two are growth processes. The third, differentiation, is
defined as the sum of chemical and morphological changes
leading to thickening and lignification of secondary cell
walls, thickening of leaf cuticle, and production of secondary
compounds (39). The essence of the Loomis concept is that
growth processes take precedence over differentiation when
conditions for cell enlargement are optimum.

In presenting the growth-differentiation concept as an
explanation of changes in host-tree secondary compounds, Lorio
and Hodges (43) present data that show dramatic increases in
solutes following the onset of drought stress conditions in the
early summer. Therefore, the osmotic adjustment concept
presented in this chapter can be viewed as a physiological
explanation of the growth-differentiation concept. Ideally,
these two concepts should be combined, because together they
offer considerable insight into source-sink dynamics and how
these dynamics influence host tree defense against herbivores.

6.6 SOUTHERN PINE DEFENSE SYSTEM

The early successional states of the southern forest
ecosystem have been extensively studied, leading to considerable understanding of the dynamics of the pine host defense
system and its interaction with attacking pine bark beetles (8,
61). Drought and sunlight competition stresses cause shifts in
the amount and chemical composition of oleoresin (7), allowing
opportunities for bark beetle colonization (55).

In southern pines, small crown ratios reduce the total
oleoresin yield (13), whereas moderate drought stress reduces
growth and stimulates oleoresin synthesis (43). Drought stress
causes a decrease in the concentration of the highly viscous
resin acids that reduce digestibility and increases the
concentration of low molecular weight monoterpenes (28).
Drought stress also reduces oleoresin exudation pressure,
although osmotic adjustment may provide a mechanism for
maintaining positive pressures during decreasing soil water
potentials. Given these complicated and opposing processes,
Sharpe and Wu (56) and Sharpe et al. (55) developed a
mathematical model that describes conifer defense against bark
beetles. The assumptions of the model were based upon the
proposed osmotic adjustment processes outlined in the present
chapter. The model describes mathematically the two types of
resin defense systems--the resin duct defense system and the
wound defense response.

The resin duct system has been identified as the primary

defense mechanism against bark beetle attack (1, 5, 7, 29, 50, 52, 53, 58). Hodges et al. (29) suggest that southern pines can be classified by their resistance or susceptibility to bark beetles using the physical properties of oleoresin such as total flow, flow rate, viscosity, and time to initial crystallization. These physical properties reflect the resin acid composition and the relative concentrations of resin acids to monoterpenes. In addition, low oleoresin exudation pressure resulting from flooding or drought is associated consistently with successful bark beetle attack (40, 41, 42, 60, 65, 66).·

There are two resin duct systems in pines, one vertical and one horizontal. They differ only in size and abundance; the vertical ducts are larger but less numerous. Annual ring duct densities of various coniferous species are believed to be affected by fluctuating weather (45), abundance of reserve carbohydrate (32, 43), mechanical injuries (2, 3, 62), temperature, sunlight, or both (23, 49), and water status (43). It was originally proposed that the resin duct system was interconnected (45), but studies using western pines in North America (59) suggest that there is only a limited amount of interchange between the duct systems. Turnover and remetabolism of the resin components are generally considered to be rapid (59) but recent studies by Lawrence (36) indicate that only monoterpenes are remetabolized. The metabolic fate of resin acids is not known.

Although the primary resin duct system has received the most attention, several investigators (5, 10, 22, 57) suggest that in the most resistant trees a combination of the resin duct system and a wound response is required to explain tree resistance or susceptibility. The wound response is an induced necrosis surrounding invasion sites by bark beetles and associated microorganisms. The wound response is believed to be a defense mechanism by which the tree stops the growth of invading fungi. The wound response does occur in southern pines, but its importance to defense against bark beetles has not been quantified (43). The wound response is also important in conifer species that do not have a primary resin system, or

in those trees where beetles survive the primary resin response
(5).

Most of the early studies on the wound response assumed
that resin and phenolics were synthesized in situ (50, 57, 72).
Recent studies by Gambliel et al. (17), however, found that the
chemical composition of wound oleoresin compounds was identical
to that synthesized in the resin duct system. The wound
response may therefore represent a method for reducing fungal
growth by the rapid "activation" of traumatic ducts in the
phloem, allowing oleoresin to saturate the infected region.

6.7 CONCLUSIONS

Osmotic adjustment has been demonstrated over a wide range
of woody plant species. It is defined as a process by which
solutes accumulate in tissues, resulting in reduced osmotic
potentials. Solute accumulation has been observed in shoots,
roots, and stems with the relative solute levels in these
organs modulated by reduction in cell enlargement and osmotic
adjustment in the phloem. Changes in cell elongation rates and
phloem loading and unloading rates are proposed as
physiological mechanisms that explain observed shifts in the
balance between growth and secondary metabolism as an element
of differentiation. The phloem osmotic adjustment hypothesis
unites the concepts of growth-differentiation (38) and seasonal
oleoresin synthesis (43). Defense against herbivores can
therefore be described in terms of the underlying adaptive
physiology of the tree.

Phloem osmotic adjustment has been proposed as a mechanism
for accumulation of solutes such as carbohydrates, amino acids,
and organic acids during drought stress. Osmotic adjustment
enhances the turgor pressure of plant tissues by raising the
solute (nutrient) content, thus increasing the suitability of
the tree as a herbivore diet. Concomitantly, osmotic adjustment provides substrates or precursors for defensive chemical
synthesis, thereby increasing the resistance of the tree to
herbivore attack. Osmotic adjustment is enhanced by environmental stress, such as moderate drought and mineral nutrient

deficiency, suggesting that it is a key physiological process modulating host-tree responses to beetle attack. Other forms of environmental stress, such as sunlight competition stress or severe drought stress, reduce osmotic concentration and thus the level of defensive chemical compounds is reduced.

ACKNOWLEDGEMENTS

The phloem osmotic adjustment hypothesis was formulated as part of research supported by NSF grant DEB 77/14406. Subsequent research has been jointly supported by US Forest Service Grant FS-30-82-4, NSF Grant BSR 84/06136, and the Texas Agricultural Experiment Station.

REFERENCES

1. Anderson, N.H. and B. Anderson. 1968. Ips bark beetle attacks and brood development on a lightning-struck pine in relation to its physiological decline. Florida Entomologist 51: 23-30.
2. Bannan, M.W. 1933. Factors in influencing the distributions of vertical resin ducts in the wood of the larch, Larix laricina (DuRoi) Koch. Proceedings, Transactions of the Royal Society of Canada. 27: 203-217.
3. Bannan, M.W. 1936. Vertical resin ducts in the secondary wood of Abietinae. New Phytologist 35: 1-46.
4. Ben-Zioni, A., Y. Vaadia and S.H. Lips. 1971. Nitrate uptake by roots as regulated by nitrate reduction products of the shoot. Physiologia Plantarum 24: 288-290.
5. Berryman, A.A. 1972. Resistance of conifers to invasion by bark beetle-fungus associations. Bioscience 22: 598-602.
6. Black, C.A. 1957. Soil Plant Relationships. pp. 116-152. Wiley, New York. Second Edition.
7. Blanche, C.A., J.D. Hodges, T.E. Nebeker and D.N. Moehring. 1983. Southern pine beetle: The host dimension. pp. 1-29. Miss. Agric. & Forestry Exp. Sta. Bull. 917, Mississippi State Univ.
8. Branham, S.J. and R.C. Thatcher, eds. 1985. Integrated Pest Management Research Symposium: The Proceedings. Tech. Rep. SO-56. U.S. Department of Agriculture, Forest Service, Southern Forest Experiment Station. Asheville, NC, April 15-18, 1985.
9. Bryant, J.P., F.S. Chapin III and D.R. Klein. 1983. Carbon/nutrient balance of boreal plants in relation to vertebrate herbivory. Oikos 40: 357-368.

10. Cates, R.G. and H.J. Alexander. 1982. Host resistance and susceptibility. In: Bark Beetles in North American Conifers: Evolution and Ecology. pp. 212-263. (J.B. Mitton and K.B. Sturgeon, eds). University of Texas Press, Austin, TX.

11. Cates, R.G., R.A. Redak and C.B. Henderson. 1983. Patterns in defensive natural product chemistry: Douglas fir and western spruce budworm interactions. In: Plant Resistance to Insects. pp. 3-19. (P. Hedin, ed). American Chemical Society, Washington, DC.

12. Chew, F.S. and J.B. Rodman. 1979. Plant resources for chemical defense. In: Herbivores: Their Interaction with Secondary Plant Metabolites. pp. 271-303. (G.A. Rosenthal and D.H. Janzen, eds). Academic Press, New York.

13. Clements, R.W. 1974. Modern Gum Naval Stores Methods. U.S.D.A. General Tech. report. SE-7. 20 p. Southeastern Forest Exp. Stn., Asheville, NC.

14. DeMichele, D.W., P.J.H. Sharpe and J.D. Goeschl. 1978. Towards the engineering of photosynthetic productivity. Critical Reviews in Bioengineering, 3: 29-92. CRC Press, West Palm Beach, FL.

15. Fares, Y., D.W. DeMichele, J.D. Goeschl and D.A. Baltuskonis. 1978. Continuously produced, high specific activity ^{11}C for studies of photosynthesis, transport and metabolism. International Journal of Applied Radiation and Isotopes. 29: 431-441.

16. Fares, Y., C.E. Magnuson and J.D. Goeschl. 1984. Maintenance of phloem turgor in water stressed plants: experimental tests using $^{11}CO_2$ extended square wave tracer kinetics. Abstract, Symposium on Carbon Transport Under Water Stress. Annual Meeting, American Society of Agronomy. Nov. 26-29, Las Vegas, NV.

17. Gambliel, H.A., R.G. Cates, M.K. Caffey-Moquin and T.D. Paine. 1985. Variation in the chemistry of loblolly pine, Pinus taeda L., in relation to infection by the blue-stain fungus, Ceratocystis minor Hedge. In: Proceedings, Integrated Pest Management Research Symposium. pp. 177-184. (S.J. Branham and R.C. Thatcher, eds.) April 15-18, 1985, Asheville, NC. Gen. Tech. Rep. SO-56. USDA, Forest Service, Southern Forest Exp. Sta. New Orleans, LA.

18. Goeschl, J.D., C.E. Magnuson, D.W. DeMichele and P.J.H. Sharpe. 1976. Concentration dependent unloading as a necessary assumption for a closed form mathematical model of osmotically driven pressure flow in phloem. Plant Physiology 58: 556-562.

19. Goeschl, J.D., C.E. Magnuson, Y. Fares and P.J.H. Sharpe. 1984. Maintenance of phloem turgor in plants under water stress: A theoretical basis for reduced photosynthetic productivity. Abstract Symposium on Carbon Transport Under Water Stress. Annual Meeting American Society of Agronomy. Nov. 26-29, 1984. Las Vegas, NV.

20. Goeschl, J.D. 1979. Hypothesis to explain the moderate level of correlation among water stress, oleoresin pressure and success of colonizing pine beetles. pp. 110-111. Proceedings 13th Annual Western Forest Insect Conference, Boise, ID. March 6-8, 1979.

21. Gosz, J. 1981. Nitrogen cycling in coniferous ecosystem. In: Nitrogen Cycling in Terrestrial Ecosystems: Processes, Ecosystem Strategies, and Management Impacts. 33: 405-426. (F. Clark and T. Roswall, eds.) Ecological Bulletins, Stockholm.

22. Hain, F.P., S.P. Cook, P.A. Matson and K.G. Wilson. 1985. Factors contributing to southern pine beetle host resistance. In: Proceedings, Integrated Pest Management Research Symposium. pp. 154-160. (S.J. Branham and R.C. Thatcher, eds.) April 15-18, 1985; Asheville, NC. Tech. Rep. SO-56. U.S. Department of Agriculture, Forest Service, Southern Forest Experiment Station. New Orleans, LA.

23. Harlow, W.M. 1931. The identification of the pines in the United States, native and introduced, by needle structure. N.Y. State Coll. Forestry Tech. Publ., 32: 21 pp.

24. Hennessey, T.C. and P.M. Dougherty. 1984. Characterization of the internal water relations of loblolly pine seedlings in response to nursery cultural treatments: Implications for reforestation success. pp. 225-243. In: Physiology and Reforestation Success. (M.L. Curyea and G.N. Brown, eds.) Martinus Nijhoff/Dr. W. Junk, Publishers, Boston.

25. Hinckley, T.M., F. Duhme, A.R. Hinckley and H. Richter. 1980. Water relations of drought hardy shrubs: osmotic potential and stomatal reactivity. Plant, Cell and Environment 3: 131-140.

26. Hodges, J.D. and P.L. Lorio. 1969. Carbohydrate and nitrogen fractions of the inner bark of loblolly pines under moisture stress. Canadian Journal of Botany 47: 1651-1657.

27. Hodges, J.D. and P.L. Lorio. 1973. Comparison of oleoresin composition in declining and healthy loblolly pines. Res. Note SO-158. U.S. Dept. Agri. For. Serv., New Orleans, LA. 4 p.

28. Hodges, J.D. and P.L. Lorio. 1975. Moisture stress and composition of xylem oleoresin in loblolly pine. Forest Science 22: 283-290.

29. Hodges, J.D., W.W. Elam, N.R. Watson and T.E. Nebeker. 1979. Oleoresin characteristics and susceptibility of four southern pines to southern pine beetle (Coleoptera: Scolytidae) attacks. Canadian Entomologist 111: 889-896.

30. Jones, M.M., N.C. Turner and C.B. Osmond. 1981. Mechanism of Drought Resistance. In: The Physiology and Biochemistry of Drought Resistance in Plants. pp. 15-37. (L.G. Paleg and D. Aspinall, eds.). Academic Press, New York.

31. Karlic, H. and J. Richter. 1983. Developmental effects on leaf water relations of two evergreen shrubs (Prunus laurocerasus L. and Ilex aquifolium L.). Flora 173: 143-150.

32. Kirsh, S. 1911. The origin and development of resin canals in the Coniferae with special reference to the development of tyloses and their correlation with the tylosal strands of Pteridophytes. Proceedings, Transactions of the Royal Society of Canada. 5: 43-83.

33. Kramer, P.J. and T.T. Kozlowski. 1979. Physiology of Woody Plants. Academic Press, New York.

34. Ku, T.T., J.M. Sweeney, and V.B. Shelburne. 1976. Preliminary evaluation of site and stand characteristics associated with southern pine beetle infestations in Arkansas. Arkansas Farm Research 25: 2.

35. Ku, T.T., J.M. Sweeney, and V.B. Shelburne. 1980. Site and stand conditions associated with southern pine beetle outbreaks in Arkansas - a hazard rating system. Southern Journal of Applied Forestry 4: 103-106.

36. Lawrence, R.H. 1971. Biosynthesis and metabolism of terpenoids in Pinus elliottii Engelm. and Pinus palustris Mill. Ph.D. Dissertation. University of Georgia, Athens, GA.

37. Levitt, J. 1980. Responses of Plants to Environmental Stresses. Academic Press, New York.

38. Loomis, W.E. 1932. Growth-differentiation balance vs. carbohydrate-nitrogen ratio. Proceedings, American Society of Horticultural Sciences 29: 240-245.

39. Loomis, W.E. 1953. Growth correlation. In: Growth and Differentiation in Plants. pp. 197-217. (W.E. Loomis, ed.) Iowa State College Press, Ames, IA.

40. Lorio, P.L. and J.D. Hodges. 1968a. Oleoresin exudation pressure and relative water content of inner bark as indicators of moisture stress in loblolly pines. Forest Science 14: 392-398.

41. Lorio, P.L. and J.D. Hodges. 1968b. Microsite effects on oleoresin exudation pressure of large loblolly pines. Ecology 49: 1207-1210.

42. Lorio, P.L. and J.D. Hodges. 1977. Tree water status affects induced southern pine beetle attacks and brood production. Res. Pap. SO-135. 7 p. U.S. Dept. Agr. For. Serv.

43. Lorio, P.L. and J.D. Hodges. 1985. Theories of interactions among bark beetles, associated microorganisms and host trees. In: U.S. Forest Service, Southern Forest Experiment Station. pp. 485-492. (E. Shoulders, ed.) Gen. Tech. Rep. SO-54. New Orleans, LA.

44. Morgan, J.M. 1984. Osmoregulation and water stress in higher plants. Annual Review of Plant Physiology. 35: 299-319.

45. Munch, E. 1919. Naturwissenschaftliche Grundlegen der Keifernharznutzung. Arbeiten aus der Biologischen Reischsanstalt für Land-und Forstwirtschaft. 10: 1-140.

46. Nilsen, E.T., P.W. Rundel and M.R. Sharifi. 1981. Summer water relations of the desert phreatophyte Prosopis glandulosa in the Sonoran Desert of southern California. Oecologia 50: 271-276.

130

47. Osonubi, O. and W.J. Davies. 1978. Solute accumulation in leaves and roots of woody plants subjected to water stress. Oecologia 32: 323-332.
48. Parker, W.C. and S.G. Pallardy. 1985. Genotypic variation in tissue water relations of leaves and roots of black walnut (Juglans nigra) seedlings. Physiologia Plantarum 64: 105-110.
49. Reid, R.W. and J.A. Watson. 1966. Sizes, distribution and number of vertical resin ducts in lodgepole pine. Canadian Journal of Botany 44: 519-525.
50. Reid, R.W., H.S. Whitney and J.A. Watson. 1967. Reactions of lodgepole pine to attack by Dendroctonus ponderosa Hopkins and blue stain fungi. Canadian Journal of Botany 45: 1115-1126.
51. Rhoades, D.F. 1979. Evolution of plant chemical defense against herbivores. In: Herbivores: Their Interaction with Secondary Plant Metabolites. pp. 3-54. (G.A. Rosenthal and D.H. Janzen, eds.) Academic Press, New York.
52. Rudinsky, J.A. 1966a. Scolytidae beetles associated with Douglas-fir: responses to terpenes. Science 152: 218-219.
53. Rudinsky, J.A. 1966b. Host selection and invasion by the Douglas-fir beetle, Dendroctonus pseudotsugae Hopkins, in coastal Douglas-fir forests. Canadian Entomologist 98: 98-111.
54. Seiler, J.R. 1984. Physiological response of loblolly pine seedlings to moisture-stress conditioning and their subsequent performance during water stress. Ph.D. Dissertation, Virginia Polytechnic and State University, Blacksburg, VA.
55. Sharpe, P.J.H., H. Wu, R.G. Cates, and J.D. Goeschl. 1985. Energetics of pine defense systems to bark beetle attack. In: Integrated Pest Management Research Symposium: The Proceedings. pp. 206-223. (S.J. Branham and R.C. Thatcher, eds.) April 15-18, 1985. Asheville, NC. Gen. Tech. Rep. SO-56. USDA, Forest Service, Southern Forest Exp. Sta., New Orleans, LA.
56. Sharpe, P.J.H. and H. Wu. 1985. A preliminary model of host susceptibility to bark beetle attack. Proceedings: International Union of Forest Research Organizations. pp. 108-127. (L. Safranyik and A.A. Berryman, eds.) Host Insect Work Group, Banff, Alberta, September, 1983.
57. Shrimpton, D.M. 1973. Extracts associated with wound response of lodgepole pine attacked by the mountain pine beetle and associated micro-organisms. Canadian Journal of Botany 51: 527-534.
58. Smith, R.H. 1975. Formula for describing effect of insect and host tree factors on resistance to western pine beetle attack. Journal of Economic Entomology 68: 841-844.
59. Stark, R.W. 1965. Recent trends in forest entomology. Annual Review of Entomology 10: 303-324.
60. Thatcher, R.C. 1960. Bark beetles affecting southern pines: a review of current knowledge. U.S. Dept. Agr. For. Serv., South. For. Exp. Stn., Occas. Pap. 180. New Orleans, LA, 25 p.

61. Thatcher, R.C., J.L. Searcy, J.E. Coster and J.D. Hertel. 1980. The Southern Pine Beetle. 247 p. USDA Forest Serv. Sci. & Ed. Admin. Tech. Bull. 1631.

62. Thomson, R.B. and H.B. Sifton. 1925. Resin canals in the Canadian spruce. Philosophical Transactions of the Royal Society of London, Ser. B., 214: 63-111.

63. Townsend, C.R. and P. Calow, eds. 1981. Physiological Ecology: An Evolutionary Approach to Resource Use. Sinauer, Sunderland, MA.

64. Tuomi, J., P. Niemela, E. Haukioja, S. Siren and S. Neuvonen. 1984. Nutrient stress: An explanation for plant anti-herbivore responses to defoliation. Oecologia 61: 208-211.

65. Vit , J.P. 1961. The influence of water supply on oleoresin exudation pressure and resistance to bark beetle attack in Pinus ponderosa. Contrib. Boyce Thompson Inst. 21: 37-66.

66. Vit , J.P. and D.L. Wood. 1961. A study of the applicability of the measurement of oleoresin exudation pressure in determining susceptibility of second growth ponderosa pine to bark beetle infestation. Contrib. Boyce Thompson Inst. 21: 67-76.

67. Wadleigh, C.H., H.G. Gauch and O.C. Magistad. 1946. Growth and rubber accumulation in guayule as conditioned by soil salinity and irrigation regime. U.S. Dept. Agr. Tech. Bull. 925. U.S. Dept. Agri., Washington, DC.

68. Waring, R.H. and G.B. Pitman. 1980. A simple model of host resistance to bark beetles. Oregon State Univ., For. Res. Lab. Res. Note 65, Corvallis, OR. 2 p.

69. White, T.C.R. 1969. An index to measure weather related induced stress of trees associated with outbreaks of Psyllids in Australia. Ecology 50: 905-909.

70. White, T.C.R. 1974. A hypothesis to explain outbreaks of looper caterpillars, with special reference to populations of Selidosema suavis in a population of Pinus radiata in New Zealnd. Oecologia 16: 279-301.

71. White, T.C.R. 1984. The abundance of invertebrate herbivores in relation to the availability of nitrogen in stressed food plants. Oecologia 63: 90-105.

72. Wong, B.L. and A.A. Berryman. 1977. Host resistance to the fir engraver beetle. 3. Lesion development and containment of infection by resistant Abies grandis inoculated with Trichosporium symbioticum. Canadian Journal of Botany 55: 2358-2365.

7. LOW TEMPERATURE: PHYSICAL ASPECTS OF FREEZING IN WOODY
PLANT XYLEM

M. F. GEORGE AND M. J. BURKE
Associate Professor, School of Forestry, Fisheries and
Wildlife, University of Missouri, Columbia, Missouri 65211,
and Associate Dean, College of Agricultural Science, Oregon
State University, Corvallis, Oregon 97331

ABSTRACT

Living cells in winter-hardy xylem of deciduous hardwoods
survive freezing by utilizing one of two methods to avoid
lethal intracellular ice formation. The first is a tolerance
of extracellular ice formation and its associated dehydration
while the second is avoidance of freezing altogether by supercooling. Extracellular freezing may be tolerated to experimental temperatures below $-196^{\circ}C$ in the most hardy species
while supercooling is limited by the ice nucleation temperature of the cell sap, usually between -40 and $-50^{\circ}C$. Ice
formation in supercooled cells occurs intracellularly and
kills the cells instantly. A simple model based on fundamental concepts of plant water relations and ice nucleation
kinetics predicts the two types of cell survival mechanisms.

7.1 EXTRACELLULAR FREEZING IN WOODY PLANT XYLEM

Most woody and herbaceous plants survive sub-freezing
temperatures by tolerating extracellular ice formation in
their tissues and the associated cellular dehydration (31).
In general, extracellular freezing may be regarded as an
equilibrium freezing process. The amount of cellular water
lost as temperature declines depends primarily on the osmotic
concentration of the cell (in higher plants the bulk of cellular water is in the vacuole) and the volumetric bulk modulus
of elasticity of the tissue (1,31,35). Cells which have lost
water to extracellular ice are under a water stress. Cells of
winter-hardy plant tissues survive this stress, the resultant

cell collapses, and the presence of extracellular ice. Tolerance of extracellular freezing in winter is species dependent
and ranges from a few ^{o}C below zero to experimental temperatures lower than $-196^{o}C$. The discussion here on extracellular
freezing will be limited to physical aspects of ice formation
in woody xylem tissues which can tolerate extreme low temperature. Mechanisms of injury during freezing will also not be
treated. The reader is referred to Levitt's treatise (31) on
temperature stress for a detailed discussion of extracellular
freezing in plants. Additionally, the reader is referred to
papers and reviews by Brown (6), Siminovitch et al. (41), and
Steponkus (42) for information on biophysical and biochemical
aspects of winter survival which are not covered here.

Sakai (38) and Sakai and Weiser (40) have shown that twigs
of many winter-hardy coniferous and deciduous species can survive immersion in liquid nitrogen ($-196^{o}C$). A freezing curve
determined by George et al. (22) for xylem from one of these
species, trembling aspen (Populus tremuloides), is shown in
Figure 1. This curve, measured by differential thermal analysis (DTA), reflects the freezing of tissue water as a function
of temperature. After ice nucleation near $-10^{o}C$, freezing
takes place in a continuous fashion. The living xylem cells
survive this freezing. It should be noted that ice nucleation
near $-10^{o}C$ is an artifact of the experiment. If ice were
present in the extracellular space as would likely be the case
under natural freezing conditions, freezing of tissue water
would occur near the melting point of the cell sap. Harrison
et al. (25) used pulsed nuclear magnetic resonance (NMR)
(proton) spectroscopy to investigate similar freezing in twigs
of red-osier dogwood (Cornus stolonifera), a hardy deciduous
shrub. Proton NMR provides a convenient tool to distinguish
between protons on liquid and solid water in frozen plant
tissues (8). It was found that tissue water in red-osier
dogwood stems froze in an analogous fashion to a dilute aqueous solution with a melting point similar to that expected for
the cell sap of the twigs. Cell wall elasticity did not

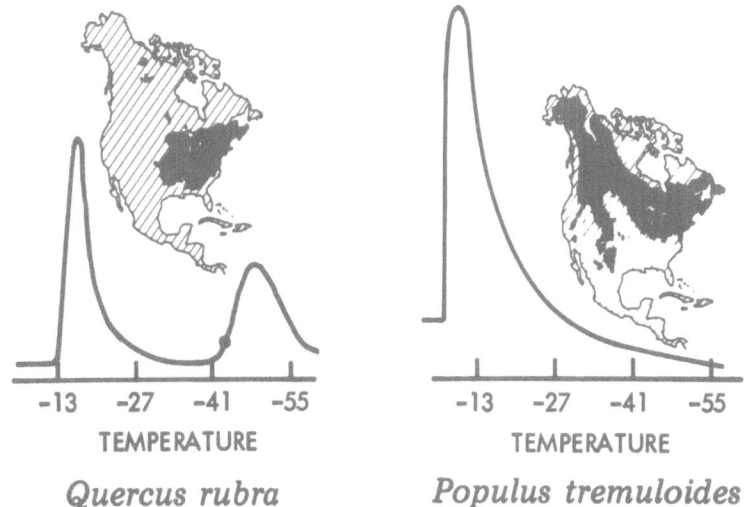

TEMPERATURE

Quercus rubra

TEMPERATURE

Populus tremuloides

FIGURE 1. DTA of winter-hardy xylem and the natural geographic distributions of northern red oak (<u>Quercus</u> <u>rubra</u>) and trembling aspen (<u>Populus</u> <u>tremuloides</u>). The freezing centered near -45°C in red oak is caused by lethal intra-cellular ice crystallization in the xylem parenchyma. High temperature freezing is extracellular and causes no injury. Aspen xylem can survive experimental exposure to -196°C.

appear to be a factor in the amount of liquid water retained at any temperature. In other words, the bulk modulus of elasticity of the tissue was small during freezing and no large tensions developed in these cells. Additionally, bound water content was found to be unrelated to survival during freezing.

The biophysics of freezing in trembling aspen and dogwood xylem is likely characteristic of most plant tissues in which cell water freezes extracellularly, regardless of the survival temperature. In this type freezing process, the osmotic properties of the cell sap play a major role in determining the amount of water frozen at any temperature. The elasticity of the cell is a secondary factor which may change the amount of cellular water remaining unfrozen at any subfreezing temperature (see below), but does not alter the general character of survival by tolerance of extracellular ice formation and its associated cell dehydration.

7.2 DEEP SUPERCOOLING IN WOODY PLANT XYLEM

Almost all woody species of temperate forest regions, numerous woody species near timberline in the Colorado Rocky Mountains, and many deciduous tree fruit cultivars avoid freezing in their xylem tissues by deep supercooling (3,20, 22). From a physical-chemical point of view, a supercooled liquid is one which is below its freezing point and is in a 'metastable equilibrium' (23). The term 'deep supercooling', as used here, refers to a fraction of supercooled tissue water in such a 'metastable equilibrium' and which freezes in a non-equilibrium fashion at temperatures ranging to almost 50°C below its melting point. When the supercooled tissue water freezes, intracellular ice crystallization occurs which kills the tissue instantly. The freezing behavior of deep super-cooled water can be easily monitored by DTA (Figure 1), NMR, or other methods.

Apple trees provide an example of deep supercooling (34). In the autumn and winter the tissues of apple trees acclimate and extracellular freezing begins near -2°C in some tissues while others deep supercool. Bark, cambial, and bud tissues freeze extracellularly and, in the most resistant cultivars, survive below -60°C. Xylem parenchyma cells in stems of apple, however, supercool and are killed by intracellular freezing associated with ice nucleations near -40°C. Hong and Sucoff (26) and Hong et al. (27) have shown that the relation-ship between freezing and xylem parenchyma injury is quantita-tive (i.e., only those cells that nucleate near -40°C are killed, adjacent cells which have not nucleated survive).

The limit of deep supercooling in xylem tissues of north-ern species often falls in the range -40 to -47°C (7,20,22, 24). This range is significant in that it covers the range where ice nucleation of the cell solutions of the tissues is expected (see below). In other words, the tissues reach their physical limit of supercooling. This limit has obvious im-plications on the latitudes at which these plants can grow. For example, the northern range of the deciduous forest of the eastern United States and southeastern Canada falls in regions where a minimum temperature of -40°C occurs at least once a year (22). Almost all angiosperm tree species native to the northern deciduous forests suffer death of some of their xylem tissues in this temperature range as a result of freezing (see Figure 1). Tree species, such as willow (Salix), paper birch (Betula papyrifera), and trembling aspen (Populus tremuloides), which are found as far north as the Arctic Circle, do not exhibit deep supercooling or at least no stable supercooling (22,24). It should be noted that boreal forest shrubs which survive under snow cover may deep supercool. The ecological implications of deep supercooling are discussed in several reviews (7,19,20).

Xylem tissues are not the only tissues of woody plants which supercool. The reader is referred to papers by Ashworth (2), Ishikawa and Sakai (28), Iwaya-Inoue and Kaku (29), Quamme (33), and Sakai (39) for discussions of supercooling in other vegetative and reproductive tissues. The reviews noted above also discuss supercooling in other plant parts.

7.3 CELLULAR ASPECTS OF DEEP SUPERCOOLING

Measurements of various physical parameters associated with the deep supercooled tissue water all support the contention that cell water is indeed supercooled (7,8,21). These include freezing-thawing hysteresis, heats of fusion, and NMR spin-lattice and spin-spin relaxation times. However, the details of the biophysical and biochemical mechanisms which allow deep supercooling are not known. Nevertheless the woody plant tissues in which deep supercooling occurs suggest that tissue structure is of primary importance.

Tissue rigidity and submicroscopic anatomical features both appear to play a role in xylem supercooling. George and Burke (21) have demonstrated that the supercooled water fraction in shagbark hickory (Carya ovata) xylem has a time constant for freezing of greater than one year at -30°C and that there are no significant kinetic barriers preventing molecular exchange of this water with water in the surrounding

environment. These results suggest that there is some lowering of free energy in the supercooled water that allows it to
come into equilibrium with ice below 0°C. The lowering of
free energy might result from large tensions (i.e., negative
hydrostatic pressures) in the supercooled solutions. Equilibrium would, therefore, be established in the system and prevent the evaporation of supercooled water and its condensation
on extracellular ice in other tissues.

The authors proposed an 'ink bottle' pore model to account
for xylem supercooling. In this model, water-filled xylem
parenchyma cells or intact xylem rays containing many parenchyma cells would be represented as large volumes. These
would be connected to extracellular air spaces or ice through
small capillary pores in the cell wall. A highly curved
interface would be established between liquid water in the
pores and extracellular vapor or ice. In a liquid-air
interface, vapor pressure equilibrium is achieved if the pores
are no larger than the maximum value defined by the Kelvin
equation for the vapor pressure above the interface. For a
liquid-ice interface, the largest pore radius allows the
melting point of pore water to be depressed sufficiently to
bring ice and water into equilibrium. This radius is defined
by considering the conditions required for mechanical
equilibrium (defined by the Laplace equation) and for physio-
chemical equilibrium (i.e., the equality of the chemical
potentials of the liquid, solid, and vapor phases) in the
system (12). A second interface would also exist in this
case. It would be between extracellular ice and vapor. A new
triple point would, therefore, be established in the system.

In addition to pore size restrictions, the materials
making up the structure of the 'ink bottle' walls (i.e., xylem
cell walls) would require a high volumetric bulk modulus of
elasticity to prevent structural collapse of the ink bottle
pores. At -35°C a tension of approximately -40 MPa would be
placed on the system. Although there are immediate questions
arising from this simple model (Do cell walls really have what

could be termed capillary pores? How can the tissues withstand the large tensions without collapse or vapor cavitation?), it is an attractive hypothesis from which to work.

If tissue structure is of paramount importance in deep
supercooling in woody plants, what role does the living pro-
toplasm play in supercooling? Obviously, living cells are
required to establish the structure necessary for supercooling
and efficient heterogeneous ice nucleators must not be present. Beyond these requirements other protoplasmic features
may or may not be of primary importance. For example, deep
supercooling is still observed in winter-hardy xylem tissues
which have been killed prior to freezing analysis; however,
the degree of supercooling is usually a few ^{o}C less (21,24).
Even more dramatically, when kiln-dried heartwood of American
sycamore (Platanus occidentalis) is saturated with distilled
water and frozen, a deep supercooled fraction of water is
observed to freeze near -36°C (18). Living xylem of American
sycamore has a supercooling peak centered at approximately
-42°C. It is clear from these observations that once the
structure of the xylem is established, supercooling is built
in; however, the degree of supercooling is influenced by the
amount of damage the living cells in the tissue have suffered.

The additional supercooling measured in living xylem is
due in part to the impact of osmotic solutes contained within
the living cells on the ice nucleation temperature of cellular
water. In mature higher plant cells most liquid water is
usually confined to the vacuole. Rasmussen and MacKenzie (37)
have shown that a 1°C depression in the melting point of a dilute aqueous solution depresses its homogeneous ice nucleation
temperature about -2°C. Pure water nucleates homogeneously
near -39°C (16). Typical melting point depressions of woody
plant xylem cells are on the order of a few ^{o}C. Since the
cell membrane property of semipermeability is required for
retention of osmotic solutes, membrane integrity is also important in deep supercooling. The volume of water freezing at
low temperature can also influence the degree of supercooling.

Volumes of micrometer dimensions depress the homogeneous nucleation temperature of water to approximately $-40^{\circ}C$ at cooling rates greater than a few $^{\circ}C/hr$. Xylem cells which deep supercool have micrometer dimensions (21), so this factor may add approximately a degree of supercooling under certain cooling conditions. It is not to be inferred that nucleation of water in supercooled plant tissues necessarily occurs in a homogeneous fashion. In fact, nucleation in shagbark hickory xylem has been shown to be heterogeneous (21).

At present, there is no evidence to suggest that cellular antifreeze agents are of importance in supercooling in xylem. Plant tissues do not generally accumulate high concentrations of the low molecular weight polyhydroxy alcohols and sugars found in certain frost sensitive insects (11,31). Melting point depressions in overwintering plants are usually between -0.5 and $-2^{\circ}C$ (31) (an osmotic concentration of 0.5 molar lowers the melting point of water approximately $-1^{\circ}C$). Similarly, there are no data suggesting that the antifreeze glycoproteins found in certain overwintering insects (11) and cold water fishes (10) are important in woody plant supercooling. Even if they are present, they would account for only a small fraction (less than $5^{\circ}C$) of the deep supercooling in woody plant tissues.

7.4 WATER RELATIONS AT LOW TEMPERATURE: A MODEL FOR FREEZING

Water in living xylem cells must be in equilibrium with extracellular water (ice) when the xylem is held at any subfreezing temperature. This is true whether a cell tolerates freezing of its water extracellularly or avoids freezing by supercooling. If as a first approximation the living xylem is envisioned as a collection of independent cells, then a simple model utilizing elementary principles of plant water relations and ice nucleation kinetics can be developed to account for either freezing survival mechanism discussed above.

The water potential of an individual cell is defined as

$$\Psi_{cell} = \Psi_{\pi} + \Psi_{\tau} + \Psi_{P}$$

where Ψ_{π} is the osmotic potential, Ψ_{τ} is the matric potential, and Ψ_{P} is the pressure potential of the cell (30). Since cell water must be in equilibrium with extracellular ice, the water potential of ice Ψ_{ice} can be substituted for Ψ_{cell}. The water potential of ice can be most simply related to temperature by solving the Clausius-Clapeyron equation assuming a constant heat of fusion. The solution gives the following:

$$\Psi_{ice} = 1.16T(^{o}C) \quad MPa$$

Alternatively, empirical relations based on measurements of the vapor pressure of water over ice and over supercooled water or refinements of the parameters of the Clausius-Clapeyron equation may be developed (see for example Fletcher (13), page 95). The relationship given above will be utilized here to eliminate unnecessary complexity since it will not alter the qualitative aspects of the model to be proposed. Additionally, Ψ_{τ} is neglected at all subfreezing temperatures considered. It will also be assumed that the cell sap behaves as an ideal solution and that Ψ_{π} can, therefore, be represented by the Van't Hoff equation

$$\Psi_{\pi} = - (RT/V)m/l_{T}$$

where R is the ideal gas constant, T is the absolute temperature, V is the molar volume of water, m is molar content of osmotically active solute in the cell, and l_{T} is the molar water content of the cell at any temperature. Finally, Ψ_{P} will be defined in its most simple form (9) as

$$\Psi_{P} = \varepsilon(l_{T}-l_{o})/l_{o}$$

where ε is the volumetric elastic modulus of the cell, l_{T} is as above, and l_{o} is the unfrozen liquid water content of the

cell (approximately 55.5 molar for a dilute cell sap). A
higher order polynomial could be incorporated here (44); however, this would again not qualitatively alter the model. The
above equations can then be combined to give the following
overall equation for cell water potential at subfreezing
temperatures:

$$1.16T(^OC) = - (RT/V)m/l_T + \varepsilon(l_T-l_o)/l_o$$

This relationship, based on elementary principles of thermodynamics and plant water relations defines the fraction of
cellular liquid water (l_T/l_o) unfrozen at any temperature.
However, it would only be specifically correct to apply this
equation by itself to model freezing for the condition that ε
is 0 at cell temperatures below the melting point of the cell
solution. If ε remains finite during freezing, tension
(i.e., negative hydrostatic pressure) will develop in the cell
and a metastable equilibrium will exist between intracellular
water and extracellular ice. The cell will be in a supercooled state. Provided that ε is sufficiently large (see below),
intracellular freezing of the cell solution will occur when
the cell temperature approaches the ice nucleation point of
the cell solution. This temperature must be predicted from a
consideration of nucleation kinetics, since the thermodynamic
basis of the overall equation for cell water potential cannot
give any information about this aspect of freezing.

The homogeneous ice nucleation rate (J) is defined as
the number of ice nuclei which form in the liquid phase in
$1/sec/m^3$ (45). Michelmore and Franks (32) have utilized the
classical nucleation theory described by Turnbull and Fisher
(43, see also Wood and Walton (45)) to express J as a function
of temperature in the following manner:

$$J = \Omega \exp(K\tau_\theta)$$

where Ω and K are constants and $\tau_\theta = 1/[\theta^3(\Delta\theta)^2]$, $\theta = T/T_m$

where T_m is the melting point of the aqueous solution in absolute temperature, and $\Delta\theta = (T_m - T)/T_m$. Values for Ω and K have been measured for a number of cell solutions including the cell sap of shagbark hickory (Carya ovata) xylem parenchyma (Table 1). Note that the parameters for hickory are considerably different than for water, suggesting that ice nucleation in xylem cells is not purely homogeneous.

TABLE 1. Nucleation constants for solutions and cells

	Ω $(1/sec/m^3)$	K	(Reference)
Water	10^{54}	1.12	(45)
saline solution	$10^{58.6}$	1.26	(17)
Erythrocytes	10^{54}	1.12	(17)
Hickory	$10^{21.4}$	0.37	(21)
Soybean (Glycine max)	$10^{14.8}$	0.07	(17)
Yeast (Saccharomyces cerevisiae)	$10^{17.9}$	0.035	(36)

If living woody xylem is considered to be a collection of a large number of independent cells, the fraction (f) of those cells which will nucleate at any particular temperature and water content (as described above) will be defined as

$$f = \exp(-Jtvl_T/l_o)$$

where t = time in sec, v = volume of an individual cell. This relation reflects the first-order kinetics of ice nucleation (17,45). In terms of the model for freezing, this quantifies the fraction of tissue water which has frozen due to ice nucleation at a particular temperature and which must be used to correct the value of l_T/l_o predicted by the water relations equation to give the actual liquid fraction.

The physical aspects of freezing in shagbark hickory xylem have been investigated in some detail (8,21) and the necessary parameters to evaluate the freezing model have been measured. The nucleation constants are as given in Table 1 and the volume of an individual xylem ray parenchyma cell is approximately $1.6 \times 10^{-14} m^3$ (21). The molarity of the cell sap (m) and the bulk modulus (ε) can then be varied to best fit the theoretical curve to the experimental measurement of tissue freezing as a function of temperature. Predicted (expectation) freezing and thawing curves utilizing the above equations with m = 2.14 molar and ε = 300 MPa are in good agreement with experimental data determined by NMR (Figure 2). These values for m and ε are not unreasonable for these cell parameters. Thawing analysis by DTA indicates that xylem cell water melts near $-3^{\circ}C$ which corresponds to an osmotic concentration of approximately 1.5 molar (8). No measurements of ε in these tissues have been attempted, but values this large are not surprising. Pressure-volume techniques have been used to measure bulk moduli in the range of 0.5 to 30 MPa in plant tissues with far less structural rigidity than hickory xylem (9). Additionally, the bulk moduli of hardwood lumber in the the radial and tangential dimensions, the weakest dimensions, are on the order of 1,000 MPa (4).

Two predicted freezing curves are shown in Figure 2. The first assumes that one cell freezes per nucleation event and the second assumes that 300 cells freeze per nucleation event (i.e., 300v replaces v in the kinetic equation). The difference in these curves indicates the weak volume dependence of nucleation. The thawing curve predicted by the model was calculated assuming an ε of 0 (see discussion below). This curve would be exactly the same as the freezing curve of a xylem tissue whose bulk modulus was 0 below the melting point of the cell solution such as red-osier dogwood xylem discussed

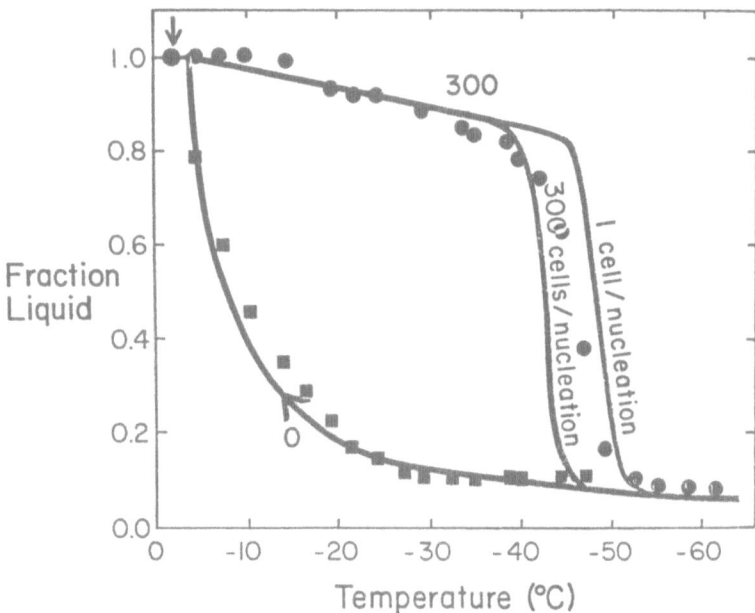

FIGURE 2. Freezing and thawing curves for winter-hardy shagbark hickory xylem. Solid lines are curves predicted by the freezing model. Closed circles and squares are actual measurements of water content during freezing made by Burke et al. using NMR (8). See text for discussion. The decrease in liquid water between -40 and -50°C as measured by NMR corresponds to the low temperature peak in a DTA experiment.

in Section 7.2. It should be noted that the model could also be used to fit freezing curves determined for tissues such as the leaves of mandarin orange(Citrus unshiu) (1) which do not demonstrate the marked freeze/thaw hysteresis of shagbark hickory xylem.

Why should freezing follow one curve with a large ε and thawing follow another curve with a small ε? The biophysical experiments discussed above (Section 7.3) and the freezing model suggest that ice nucleation and intracellular freezing begin to take place near -40°C and are complete by -50°C. Once ice is within a xylem cell, a new equilibrium is estab-

lished between the solid and liquid phases within the cell.
No hydrostatic pressure difference will exist between intracellular ice and liquid in this new state and when cellular
temperature subsequently is raised, thawing will proceed in
the fashion expected for a simple aqueous solution as predicted by the relationship between Ψ_{ice} and the Van't Hoff
equation.

An inherent problem of the model is that its validity depends on the concept of large negative hydrostatic pressures
in plant cells. This concept not only requires a high bulk
modulus, but also requires that cavitation be avoided. For
example, in shagbark hickory xylem at $-30^{\circ}C$, the calculated
hydrostatic pressure is -30 MPa. In the classic work on water
cavitation by Briggs (5), the limiting hydrostatic pressure of
water in glass capillaries only approached -30 MPa at $+6^{\circ}C$.
On cooling to $0^{\circ}C$, a 10 fold reduction in the limiting
hydrostatic pressure was observed. Briggs (5) did not make
measurements on supercooled systems. If the limiting
hydrostatic pressure were only -3 MPa for supercooled
solutions, cells in the shagbark hickory xylem described here
would have cavitated and frozen at temperatures far above
$-40^{\circ}C$. It is difficult to reconcile this unless it is assumed
that Brigg's anomaly at $0^{\circ}C$ was an experimental artifact, was
an interfacial property specific to water and glass, or was
limited to the temperature region near and slightly above $0^{\circ}C$.
The concept of large negative hydrostatic pressures in plant
cells is attractive because it allows a very simple model
which can account for the maintenance of a metastable equilibrium between cell solutions and extracellular ice in systems
such as shagbark hickory xylem.

An additional deficiency of the proposed model is that it
does not readily account for low temperature freezing events
at higher temperatures. For example, during winter acclimation, xylem supercooling in shagbark hickory (8) shifts from
near $-10^{\circ}C$ in early autumn to $-45^{\circ}C$ by winter. Heterogeneous
ice nucleation parameters as described by Fletcher (13-15)

would have to be incorporated to account for this intermediate supercooling.

7.5 SUMMARY

Biophysical measurements of freezing in woody plant xylem indicate that there are two diverse ways by which living cells of different species avoid lethal intracellular ice crystallization. The first method is by tolerating extracellular ice formation and its associated cellular dehydration. In many tree species this may be accomplished to experimental temperatures as low as $-196^\circ C$. The second method is one by which intracellular water supercools and is not lost to ice in surrounding tissues. Both of these methods of cell survival involve an equilibrium between intracellular water and extracellular ice; however, the supercooled state is a metastable state and has a physical limit which is imposed by the ice nucleation temperature of the cell sap. This limit is generally between -40 and $-50^\circ C$ and has consequences on the natural distributions of those tree species that utilize supercooling as a survival mechanism. Despite the diverse nature of the two survival mechanisms, a simple freezing model based on plant water relations and ice nucleation kinetics can account for both. The model suggests that the volumetric bulk modulus of elasticity of the cells which make up the xylem will determine which mechanism is displayed.

Note: the algorithms necessary to generate solutions for the freezing model are contained in an IBM PC BASICA program which is available upon request from M. J. Burke. A similar program written in Technical Systems Consultants, Inc. XBASIC which runs under the FLEX operating system on 6809 microcomputers is available from M. F. George.

REFERENCES

1. Anderson, J. A., L. V. Gusta, D. W. Buchanan, and M. J. Burke. 1983. Freezing of water in citrus leaves. J. Amer. Soc. Hort. Sci. 108(8): 397-400.

148

2. Ashworth, E. N. 1982. Properties of peach flower buds which facilitate supercooling. Plant Physiol. 70: 1475-1479.
3. Becwar, M. R., C. Rajashekar, K. J. Hansen Bristow, and M. J. Burke. 1981. Deep undercooling of tissue water and winter hardiness limitations in timberline flora. Plant Physiol. 68: 111-114.
4. Bodig, J., and B. A. Jayne. 1982. Mechanics of Wood and Wood Composites. Van Nostrand Reinhold Company, N. Y. 712 p.
5. Briggs, L. J. 1950. Limiting negative pressure of water. J. App. Phys. 21: 721-722.
6. Brown, G. N. 1978. Protein synthesis mechanisms relative to cold hardiness. Pages 153-163. In: Plant Cold Hardiness and Freezing Stress: Mechanisms and Crop Implications (P. H. Li and A. Sakai, eds.). Academic Press, N. Y.
7. Burke, M. J., L. V. Gusta, H. A. Quamme, C. J. Weiser, and P. H. Li. 1976. Freezing and injury in plants. Ann. Rev. Plant Physiol. 27: 507-528.
8. Burke, M. J., M. F. George, and R. G. Bryant. 1975. Water in plant tissues and frost hardiness. Pages 111-135. In: Water Relations of Foods (R. B. Duckworth, ed.). Academic Press, N. Y.
9. Dainty, J. 1976. Water relations of plant cells. Pages 12-35. In: Encyclopedia of Plant Physiology, New Series, Volume 2, Part A: Transport in Plants II, Part A, Cells (U. Luttge and M. G. Pitman, eds.). Springer-Verlag, Berlin.
10. Devries, A. L. 1982. Biological antifreeze agents in coldwater fishes. Comp. Biochem. Physiol. Vol. 73A, No.4: 627-640.
11. Duman, J. G. 1982. Insect antifreezes and ice-nucleating agents. Cryobiology 19: 613-627.
12. Everett, D. H. 1961. Thermodynamics of frost damage to porous solids. Trans. of the Faraday Soc. 57(9): 1541-1551.
13. Fletcher, N. H. 1970. The Chemical Physics of Ice. Cambridge Univ. Press, London. 271 p.
14. Fletcher, N. H. 1963. Nucleation by crystalline particles. J. Chem. Phys. 38: 237-240.
15. Fletcher, N. H. 1959. Size effect on heterogeneous nucleation. J. Chem. Phys. 29: 572-576.
16. Franks, F. 1982. The properties of aqueous solutions at subzero temperatures. Pages 215-338. In: Water: A Comprehensive Treatise (F. Franks, ed.). Plenum Press, N. Y.
17. Franks, F., S. F. Mathias, P. Galfre, S. D. Webster, and D. Brown. 1983. Ice nucleation and freezing in undercooled cells. Cryobiology 20: 298-309.

18. George, M. F. 1983. Freezing avoidance by deep supercooling in woody plant xylem: preliminary data on the
importance of cell wall porosity. Pages 84-95. In:
Current Topics in Plant Biochemistry and Physiology, Vol.
2 (D. D. Randall, D. G. Blevins, R. L. Larson, and B. J.
Rapp eds.). Univ. of Missouri Press, Columbia.

19. George, M. F., M. R. Becwar, and M. J. Burke. 1982.
Freezing avoidance by deep undercooling of tissue water
in winter-hardy plants. Cryobiology 19: 628-639.

20. George, M. F., and M. J. Burke. 1980. The occurrence of
deep supercooling in cold hardy plants. Pages 1-14. In:
Commentaries in Plant Science, Volume 2 (H. Smith, ed.).
Pergamon Press, N. Y.

21. George, M. F., and M. J. Burke. 1977. Cold hardiness
and deep supercooling in xylem of shagbark hickory.
Plant Physiol. 59: 319-325.

22. George, M. F., M. J. Burke, H, M. Pellett, and A. G.
Johnson. 1974. Low temperature exotherms and woody
plant distribution. HortScience 9(6): 519-522.

23. Glasstone, S. 1946. Textbook of Physical Chemistry, 2nd
Edn. D. Van Nostrand, N. Y.

24. Gusta, L. V., N. J. Tyler, T. H. Chen. 1983. Deep
undercooling in woody taxa north of the $-40^{\circ}C$ isotherm.
Plant Physiol. 72: 122-128.

25. Harrison, L. C., C. J. Weiser, and M. J. Burke. 1978.
Freezing of water in red-osier dogwood stems in relation
to cold hardiness. Plant Physiol. 62: 899-901.

26. Hong, S., and E. Sucoff. 1980. Units of freezing of
deep supercooled water in woody xylem. Plant Physiol.
66: 40-45.

27. Hong, S., E. Sucoff, and O. Y. Lee-Stadelmann. 1980.
Effect of freezing deep supercooled water on the viabil-
ity of ray cells. Bot. Gaz. 141: 464-468.

28. Ishikawa, M., and A. Sakai. 1982. Characteristics of
freezing avoidance in comparison with freezing tolerance:
a demonstration of extraorgan freezing. Pages 325-340.
In: Plant Cold Hardiness and Freezing Stress: Mechanisms
and Crop Implications, Volume 2 (P. H. Li and A. Sakai,
eds.). Academic Press, N. Y.

29. Iwaya-Inoue, M., and S. Kaku. 1983. Cold hardiness in
various organs and tissues of Rhododendron species and
the supercooling ability of flower buds as the most
susceptible organ. Cryobiology 20: 310-317.

30. Kramer, P. J. 1969. Plant and Soil Water Relationships:
A Modern Synthesis. McGraw-Hill Book Co., N. Y. 482 p.

31. Levitt, J. 1980. Responses of Plants to Environmental
Stresses: Chilling, Freezing and High Temperature Stresses, 2nd Edn., Volume 1. Academic Press, N. Y. 497 p.

32. Michelmore, R. W., and F. Franks. 1982. Nucleation
rates of ice in undercooled water and aqueous solutions
of polyethylene glycol. Cryobiology 19: 163-171.

33. Quamme, H. A. 1978. Mechanism of supercooling in over-wintering peach flower buds. J. Amer. Soc. Hort. Sci. 103(1): 57-61.

34. Quamme, H. A., C. Stushnoff, and C. J. Weiser. 1972. The relationship of exotherms to cold injury in apple stem tissues. J. Amer. Soc. Hort. Sci. 97(5): 608-613.

35. Rajashekar, C., and M. J. Burke. 1982. Liquid water during slow freezing based on cell water relations and limited experimental testing. Pages 211-219. In: Plant Cold Hardiness and Freezing Stress: Mechanisms and Crop Implications, Volume 2 (P. H. Li and A. Sakai, eds.). Academic Press, N. Y.

36. Rasmussen, D. H., M. N. Macaulay, and A. P. Mackenzie. 1975. Supercooling and nucleation of ice in single cells. Cryobiology 12: 328-329.

37. Rasmussen, D. H., and A. P. Mackenzie. 1972. Effect of solute on ice-solution interfacial free energy: calculation from measured homogeneous nucleation temperatures. Pages 126-145. In: Water Structure at the Water-Polymer Interface (H. H. G. Jellinek, ed.). Plenum Press, N. Y.

38. Sakai, A. 1960. Survival of twigs of woody plants at -196°C. Nature (London) 185: 393-394.

39. Sakai, A. 1979. Freezing avoidance mechanism of primordial shoots of conifer buds. Plant Cell Physiol. 20: 1381-1390.

40. Sakai, A., and C. J. Weiser. 1973. Freezing resistance of trees in North America with reference to tree regions. Ecology 54(1): 118-126.

41. Siminovitch, D., J. Singh, and I. A. De La Roche. 1975. Studies on membranes in plant cells resistant to freezing. I. Augmentation of phospholipids and membrane substance without changes in unsaturation of fatty acids during hardening of black locust bark. Cryobiology 12: 144-153.

42. Steponkus, P. L. 1984. Role of the plasma membrane in freezing injury and cold acclimation. Ann. Rev. Plant Physiol. 35: 543-584.

43. Turnbull, D., and J. C. Fisher. 1949. Rate of nucleation in condensed systems. J. Chem. Phys. 17: 71-73.

44. Tyree, M. T., and H. T. Hammel. 1972. The measurement of the turgor pressure and the water relations of plants by the pressure-bomb technique. J. of Exp. Bot. 23(74): 267-282.

45. Wood, G. R., and A. G. Walton. 1970. Homogeneous nucleation kinetics of ice from water. J. of App. Phys. 41(7): 3027-3036.

8. MULTIPLE STRESS FACTORS: THE POTENTIAL ROLE OF SYSTEM MODELS IN ASSESSING THE IMPACT OF MULTIPLE STRESSES ON FOREST PRODUCTIVITY

R.L. GRAHAM, T.R. FOX AND P.M. DOUGHERTY

Forest Ecologist, Weyerhaeuser Company, Centralia Research Center, Centralia, Washington 98531; Soil Scientist, Weyerhaeuser Company, Centralia Research Center, Centralia Washington 98531; and Manager, Weyerhaeuser Company Forestry Research Field Station, Wright City, Oklahoma 74766

ABSTRACT

Groundrules, advantages, and disadvantages to using systems modeling to address stress management questions are discussed. A systems model designed to explore the effects of multiple climatic stresses on potential site productivity is presented as an example of systems modeling applied to stress management research. Predictions from this model are used to examine the effect of multiple versus single climatic stresses on seasonal stand productivity. The model is also used to explore the possible effect of changing climates on future loblolly pine plantation productivity.

8.1 INTRODUCTION

Forest ecosystems frequently experience multiple environmental stresses simultaneously. Excessive air and soil temperatures commonly occur in association with low soil moisture and low humidity. Low light may be associated with cold air temperatures. Insect attacks may be exacerbated by drought conditions. Furthermore, a forest generally exhibits multiple responses to a single stress. For instance, warm temperatures may enhance root growth but increase respiration. To successfully manage forests to mitigate or avoid stress, a holistic approach is required. We must synthesize our knowledge about individual stresses and tree response to be able to predict the net response of a forest to a set of conditions.

Systems modeling is one tool, which can aid both forest scientists and managers in integrating their knowledge about tree physiology and environment, thereby improving their ability to manage forests. In this paper, we will first present some groundrules for modeling and discuss the advantages and disadvantages of using systems modeling as a tool in stress management. Secondly, we will illustrate how systems modeling can and might be used.

8.2 GROUNDRULES FOR SYSTEMS MODELING

8.2.1 Rule 1. Define specific objectives for the modeling project.

Many modeling projects fail because their objectives were poorly defined. The first step in any modeling project should be to define both the analytical and organizational objectives of the model. Analytical objectives define what the model will mathematically predict and to what resolution. Organizational objectives describe how and for what the model will be used and who will use it. The following questions should be answered in developing these objectives: Is the model intended to be a management tool? Is it intended to be a teaching tool? Who will actually run the model if it's a computer model? Who will use the predictions generated by the model? How will the predictions be used? What is the model's expected lifetime? With these questions answered at the beginning, the project is far more likely to succeed.

8.2.2. Rule 2. Develop criteria for judging the model.

If a model is valid, then it has met its analytical objectives. If a model is useful, then it has met its organizational objectives. Criteria for assessing both validity and usefulness should be developed at the same time a model's objectives are defined; that is, before the model is built. Often in defining objectives and criteria, the scope of the modeling effort is altered because it becomes clear that data or resource limitations will preclude a successful completion of the project in its original form. Having well

defined criteria for assessing the model at the start of the
modeling project will help steer the project along a pathway
that will ensure its success. These criteria are particularly
important for models intended for management rather than
research purposes.

8.2.3 Rule 3. Recognize the limitations of specific models and modeling in general.

Data, or the lack thereof, often constrains a model's
usefulness. It is essential to recognize when a model is describing hypotheses and assumptions rather than actual data
about a system. For instance, at present there is very little
data on root growth in forest plantations and how it varies
with stress, age, or management practices. Thus any model
which depends on describing root growth dynamics is necessarily somewhat hypothetical. Whereas a model which depends
on describing photosynthesis may be more factual as there is a
good database available. Modeling is not a substitute for
knowledge. Modeling at best can only complement knowledge.
In any research program, modelling must go hand in hand with
data collection and experimentation.

8.3 ADVANTAGES AND DISADVANTAGES TO SYSTEM MODELING

8.3.1 Advantages

The foremost advantage of systems modeling is information
syntheses. A good model brings together all the relevant facts
about a system and organizes them in a logical and scientifi-
cally valid manner. The model allows examination of a system
as a whole rather than fragments.

Because a model quantifies the hypotheses within it, a
model provides a vehicle for scrutinizing and testing
assumptions about a system. If a model's predictions are
proven wrong then a hypothesis within the model must be invalid.

A model may also help guide and rank future research
priorities. A model can do this in two ways. First, the
process of building the model forces the builder to organize
facts and hypotheses about the system and to form links among
them. This process can indicate where data is lacking and

identify where further research is needed. Second, the model
may make counterintuitive predictions which demand further
experimentation to prove or disprove.

Finally, models can be useful tools for management
decision-making. They provide a means of passing on understanding or "wisdom" about a complex system. They can also
quickly show managers what is known and perhaps more importantly what is not known about a system. If a model is intended as a management tool, then it's advisable to include the
managers in the model's creation. By including the managers,
confidence and familiarity with the model is developed and the
model is more likely to be used and used correctly.

8.3.2 Disadvantages

Models reflect their builder's conceptions and biases
about a system. Thus, they will generally predict what the
builder thought they would predict before ever having built the
model, particularly when assumptions concerning missing data
are made. The model will probably predict that the world runs
just like the builder thought it ran. The danger in this is
that the builder or user may accept the model's predictions as
verification of the original hypotheses. Data verify hypothesis, models don't.

The inherent biases of a model restricts its usefulness.
For instance, two of the better known forest ecosystem models,
FORCYTE and FORTNITE, designed to examine the effect of repeated harvests on long term forest productivity are predicated on
the idea that nutrient availability limits forest productivity
(1,2,3,4,15). They are inherently insensitive to water limitations and thus their usefulness is diminished when dealing
with water stressed ecosystems and harvesting practices which
affect water availability.

There is often a tendency to try to include everything
that is known about a system when building a systems model.
This occurs in part because the objectives and criteria for
judging the model are poorly defined and in part because the
builder wishes to avoid criticism. Modelers are often faulted

for leaving something out. The dangers in building all-inclus-
ive models are many. First the model may never be finished
and ready for testing because there are always ways to
"improve" the model and new information to add. Second, such
models are difficult to test because they are too large. If
they are tested and errors in prediction are determined to
exist, it may be difficult to backtrack and determine what
aspect of the model is producing the erroneous prediction. The
time and effort spent on developing and testing models that
attempt to be all inclusive but are heavily dependent on
assumptions would generally be much better spent on experimentation and data collection.

Finally, even when a modeling project is well designed and
the model fulfills its intended objectives, there is always the
risk that the model will be used inappropriately. A model may
be pushed to make predictions outside its analytical boundaries.
For instance, a model designed for natural forests might be
used to predict the growth response of managed plantations.
Models which are designed to be strategic planning tools may be
used erroneously to make short term, site specific decisions.

8.4 EXAMPLES OF MODELS IN FORESTRY

In spite of the limitations, systems modeling has been
successfully used in stress management and stress management
research. Most of the actual management models have been
concerned with insect outbreaks, for example, the Douglas-fir
tussock moth population outbreak model built by the U.S. Forest
Service (7). We presume the emphasis on insect outbreaks stems
from two factors. First, there are tools such as insecticides,
which can be used to directly combat these stresses; whereas
for other stresses such as excessive heat, or drought, one is
generally limited to indirect methods for controlling the
stress. Secondly, there are better databases for building
reliable insect models than there are for some of the other
stresses a forest experiences.

Several research-oriented models concerning water or
nutrient stresses have also been published and are currently

being used as teaching or research tools. There are at least
two water stress models for forests of the Rocky Mountains and
a detailed water uptake and transpiration model for Cascade
Douglas-fir forests (13,14,25,26). FORCYTE, and FORTNITE, are
two forest ecosystem models designed to describe the relationship between nutrient cycling and forest productivity (1,2,3,
4,15). Both models have been used as research and teaching
tools to examine the impact of intensive harvesting on long-
term forest productivity.

Much good research modeling has been done outside the
United States. There are models of water uptake and tran -
spiration for coniferous forests of the U.K. (12,20). Forest
nutrient dynamics have been modeled for Swedish coniferous
forests (5,6). The Dutch pioneered modeling agronomic ecosystems and are now modeling forest ecosystems (8,19). In
Finland canopy light interception has been modeled and predictions made of the relative amount of inter and intra tree
shading (21,22,23,23). South of the equator there are several
productivity models of radiata pine (9,16,17,18).

8.5 USE OF A MODEL IN STRESS MANAGEMENT RESEARCH

Although most of these modeling efforts have not had
stress management as a specific objective, they have been useful in enhancing our understanding about forest ecosystems.
At this point we would like to demonstrate how modeling can be
used as a tool in stress management. Our desire here is not to
convince the reader that this specific model is correct but
rather to show what possibilities modeling offers to stress
management.

8.5.1. Model description

The model we have selected to use is an ecophysiological
model of forest productivity developed at Weyerhaeuser Company.
The analytic objective of the model is to calculate the
potential productivity of a site based on the site environment
and basic physiological processes. The organizational objective of building the model was to identify research strategies

and experiments which would yield information that could be used to manipulate forest productivity. The model is not intended to give quantitative site specific predictions but rather to yield predictions which can be used to make relative comparisons between different environmental scenarios. It was the builder's belief that our knowledge of forest physiological processes and their relationship to the environment was too imperfect to be able to build quantitatively accurate models of forest productivity.

The model is relatively simple in structure and has a limited number of parameters. Total incoming net radiation, wind speed above the canopy, air temperature, saturation deficit of the atmosphere, precipitation, soil moisture holding capacity, and the amount of nitrogen actively cycled on the site define the environment. The timestep is one day and model is intended to be run for one year. The model predicts total net productivity based on the balance between gross stand photosynthesis and total respiration. The model does not consider the allocation of carbohydrates or stand structure. A complete description of the model equations and parameters is found in Graham et al. (11).

8.5.2 Using model to integrate the effects of multiple stresses

In our first example the model is used to examine the relationship between climate and seasonal stand productivity. Our purpose in this exercise was to develop a better understanding of the way climatic factors affect productivity when considered together or individually, given the logic of our model. To do this we first set all the model stand parameters to simulate a productive middle latitude coniferous forest. The stand parameters are not typical of loblolly pine or Douglas-fir but instead represent a hypothetical average tree. We next defined the optimal environment for growth of these trees as one with a year-round air temperature of $15^{O}C$, a daily average net radiation of 280 watts/m^2, a daily precipitation of 10mm, and 100% relative humidity. The soil water holding capacity was set to 300mm and nitrogen was assumed to never be limiting. We then selected two locations, Seattle, Washington and Fort Smith, Arkansas for which we could obtain daily climatic data for a "typical year" (Table 1). Seattle has a maritime climate with a regular summer drought and winter rains. Fort Smith, Arkansas has a continental climate with colder winters and much warmer summers than Seattle. Precipitation is distributed over the year but periodic droughts are common.

TABLE 1. Monthly summaries of environmental test data sets.

	Seattle, Washington					Fort Smith, Arkansas				
Month	Temp	Par	Wind	Rh	Ppt	Temp	Par	Wind	Rh	Ppt
	(oC)	(E/m²* day)	(m/S)	(%)	(mm/ mth)	(oC)	(E/m²* day)	(m/s)	(%)	(mm/ mth)
Jan	4.2	7.9	4.3	78	135	1.5	17.3	4.1	64	105
Feb	5.8	10.3	4.0	66	192	4.4	25.8	3.3	69	33
Mar	6.9	19.1	3.6	67	94	12.6	31.4	3.8	57	50
Apr	8.8	33.4	4.5	55	52	13.9	42.5	3.8	58	32
May	12.8	42.9	3.6	56	16	21.5	41.0	3.1	70	132
Jun	17.6	46.6	3.3	54	26	23.1	54.6	2.6	71	101
Jul	17.3	41.8	3.5	62	14	27.4	47.6	2.4	68	96
Aug	18.7	31.2	3.5	61	16	27.7	47.0	2.5	70	59
Sep	16.1	26.1	3.2	72	37	22.6	39.8	2.4	62	16
Oct	11.7	15.0	3.7	76	103	17.8	33.9	2.6	62	89
Nov	6.4	8.2	3.7	74	134	10.3	14.3	3.6	70	145
Dec	5.0	4.0	3.8	82	174	7.2	18.2	3.6	73	152
Year	11.0	23.9	3.7	67	998	15.8	34.5	3.2	66	1015

*E - moles of photons

For each location, we predicted productivity of this "generic" forest four times. Each time one of the climatic variables considered in the model took on the actual daily site values while the other climatic variables were kept at our predetermined "optimum." Figures 1a & b show the results of the four runs in which one climatic variable took the actual daily site values while the others were "optimal." Examinations of the trends in productivity predicted in this manner permits an assessment of when and how the climatic factors considered (RH, Ppt, Temp, PAR) might limit productivity.

Relative humidity and wind - Predicted impact on productivity:

The model predicts that the observed pattern of relative humidity and wind for a typical year at Seattle or Fort Smith would not have a major effect on productivity if temperature, radiation, and precipitation were optimized. This result is unrealistic because temperature, radiation, and precipitation and thus soil water potential are all optimum. In the field this situation would not normally occur. Also, because the daily relative humidity value used in our model is the average value for the period between 7 a.m. and 7 p.m., the values for any one day would not be exceedingly low. Direct effects of vapor pressure deficit on stomata have in fact been reported and would be expected to periodically limit productivity.

Temperature - Predicted impact on productivity:

The predicted response to a "typical year's" temperatures with all the other climatic variables optimized differed between Seattle and Fort Smith. Under the Seattle "typical year" climate, temperature is predicted to decrease productivity below the maximum line only in the late fall and winter season. Even during this period the model predicts that productivity would be 50% of maximum.

Under "typical year" conditions for Fort Smith, when all climatic variables except temperatures are optimized, the model predicts a winter depression of slightly greater magnitude

160

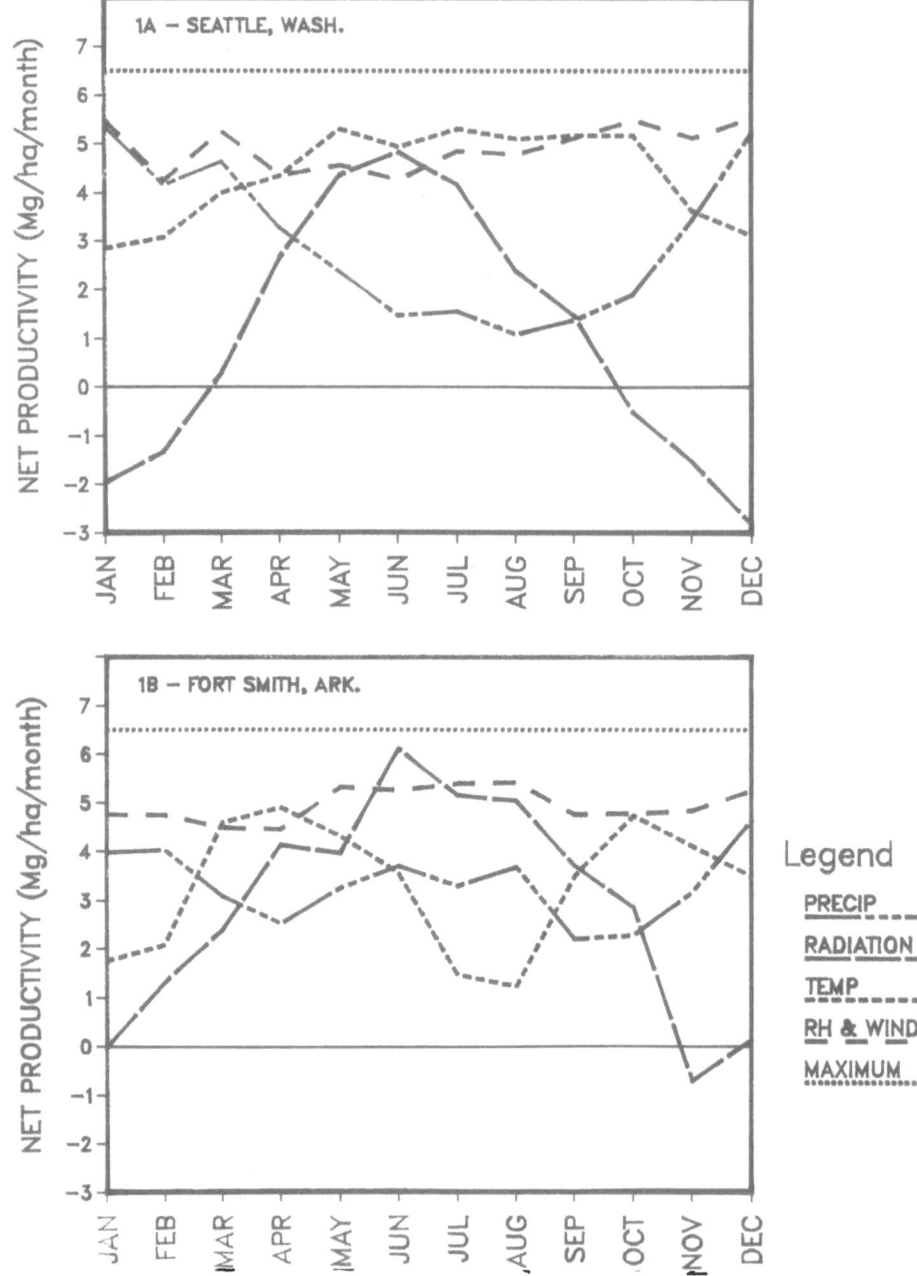

FIGURE 1. Effect of an individual climatic variable on net productivity at a) Seattle, WA and b) Fort Smith, AR when all other climatic variables are optimum. The maximum line reflects productivity when all variables are at optimum.

than that predicted for the Seattle area. In addition, at Fort Smith a large depression in net productivity is predicted to occur in July and August. This is surprising because most reports suggest that soil water is the major limiting variable in this area during the summer. The model predicts that even if adequate soil water was available, summer productivity would still be severely reduced due to increased respiration.

Radiation - Predicted impact on productivity :

The model predicts that if radiation is allowed to vary according to the daily values observed for a "typical year" in Seattle, while other climatic variables are optimized, radiation would severely limit productivity in the winter period and moderately limit productivity in the spring and fall. In fact net productivity is predicted to be negative for most of the winter (respiration exceeds photosynthesis). Of course, this depression is much greater than would typically occur because temperature is held at a constant high value and consequently respiration rates are higher than would naturally occur. For similar model conditions at Fort Smith, radiation is predicted to be much less limiting, although productivity is predicted to be near zero for November-February. Unlike Seattle, productivity at Fort Smith is expected to be only slightly negative even though daily air temperature is held at 15°C and respiration is running at a much higher rate than actually would occur in nature.

Precipitation - Predicted impact on productivity :

When precipitation was allowed to take on the values observed for a typical year in Seattle and the other climatic variables optimized, the model predicted that productivity would be severely limited during the late spring and summer period. The model suggests that precipitation is limiting productivity in the Seattle area more than is commonly supposed.

For similar conditions at Fort Smith, precipitation was predicted to have a much less depressing effect on productivity than what was predicted for Seattle. No major summer depres-

sion was predicted for the "typical Fort Smith year". This
would not be true for droughty years which do occur.

Climatic variables combined - Predicted impact on productivity:
    Figures 2a and b illustrate the model's predictions con-
sidering each climate factor independently, as discussed in
the previous sections, and the predicted trend in productivity
if all the climatic factors (radiation, temperature, precipitation, relative humidity and wind) are assigned their actual
site climate values. The hatched area shown on figures 2a and
b reflects the difference in productivity when productivity is
controlled by all the climatic factors simultaneously versus
when productivity is controlled by the single most limiting
climatic variable.

    For the Seattle area, using the most limiting variable to
describe productivity over the year yields about the same predicted trend as when all the variables act together to determine productivity - the main exception being in the winter
period. During this period the prediction for the single most
limiting variable (radiation) indicated negative productivity
because temperature was held at 15°C. With all the climatic
factors taking on their observed values, the magnitude of negative productivity predicted for the winter period was less
than that predicted by the most limiting variable. The fact
that a period of substantial negative productivity is predicted
for the Seattle area for a "typical year" in either case suggests that respiration losses during this period when photosynthesis is limited by low radiation would be important in
determining productivity in this area. This has also been
mentioned by Emmingham and Waring (10).

    The contrast between the productivity predictions for
Fort Smith using the single most limiting climatic variable
versus the predictions using the combined simultaneous variables suggests that productivity at Fort Smith may be much more
dependent on interacting stress responses than at Seattle (Fig
2b). Throughout most of the year predicted productivity is
less when predicted for multiple variable conditions than when

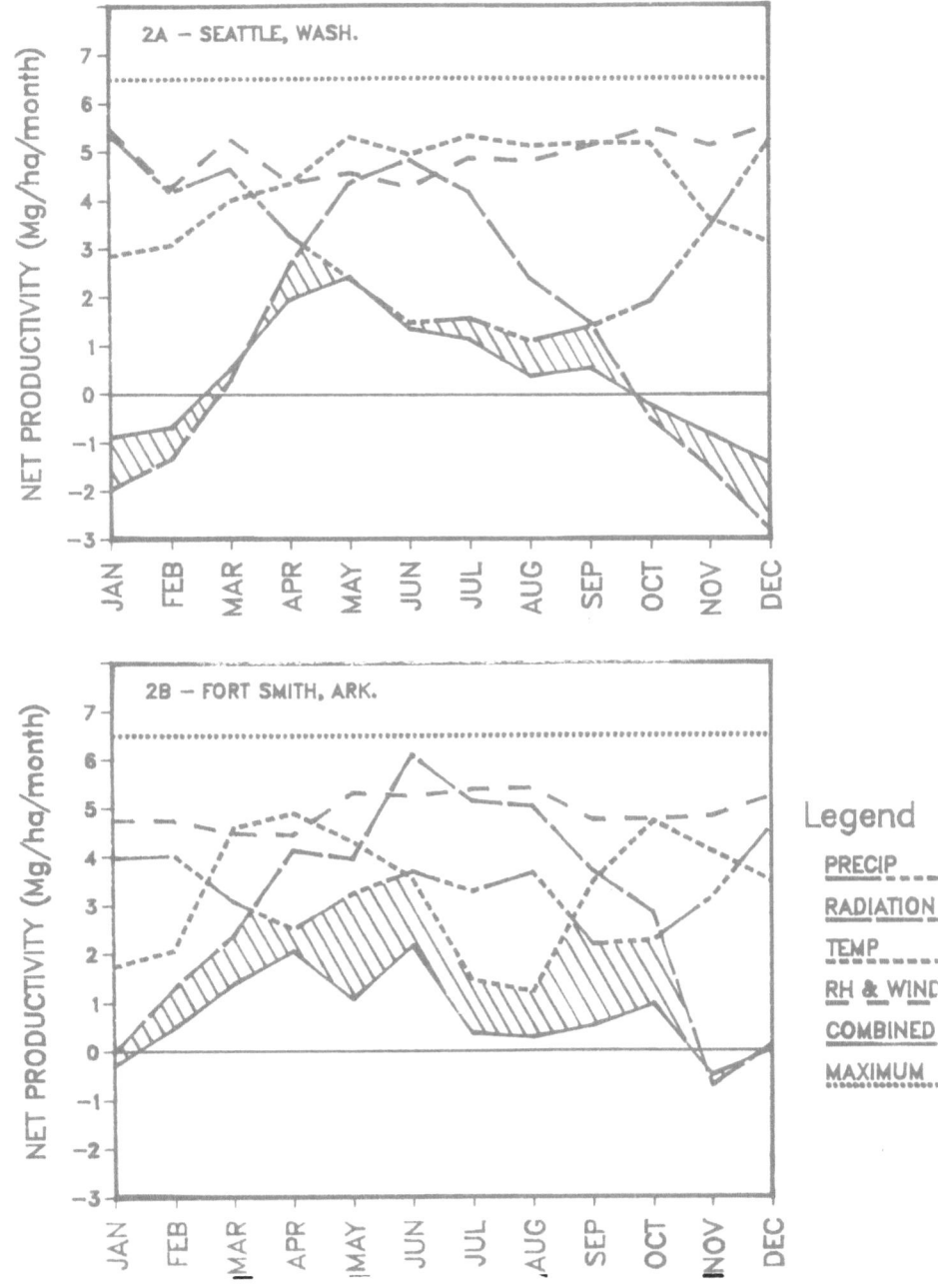

FIGURE 2. Comparison of the effect of all climatic variables combined vs. individual climatic variables on net productivity at a) Seattle, WA and b) Fort Smith, AR

the single most limiting variable is used to predict productivity. This suggests that no one climatic variable is controlling productivity at any time. More research is needed
to quantify tree responses to multiple climatic stresses.

8.5.2.1 <u>Research needs</u> - The derivation of the model used
in the example we have discussed was not presented in this
paper but has been described in detail elsewhere (11). During
the development of this model it quickly became apparent that
additional information was needed to describe the relationship
between environmental factors and net productivity. Some of
the specific areas where additional data were needed to improve the predictive capability of the model were
     Leaf area dynamics
     Fine root turnover and nutrient translocation rates
     Temperature effects on:
       - Respiration
       - Photosynthesis
     Decomposition rates
     Solar radiation data
     Physiological measurements on whole trees in real systems

8.5.3 <u>Using model to explore the effect of climatic change</u>
     In the previous example, we illustrated how systems
modeling could be used to explore the effect of combined
stresses, in this case climatic stresses, on a forest para-
meter such as net productivity. In the next example, we
illustrate how a systems model might be used to help predict
the effect of stresses which cannot easily be stimulated in a
controlled experiment.
     Our objective in this example was to examine the effects
of 1) a drought and 2) a climatic shift due to increased global
CO_2 level on the productivity of a poor loblolly pine site and
a good loblolly pine site in Arkansas. To do this we set the
parameters of the previously defined model to represent a good
or poor loblolly site. The good site had a waterholding capacity of 300 mm, a projected leaf area index of $3.5m^2/m^2$, and

$40g/m^2$ of nitrogen actively being cycled. The poor site had a waterholding capacity of 50mm, a projected leaf area index of $2.25m^2/m^2$, and $30g/m^2$ of nitrogen. We then ran the model for both sites using the climatic data for (1) a typical year at Fort Smith, Arkansas, (2) a year in which rainfall was 50% below normal between April and September (drought year) and (3) a year in which the precipitation was 20% less than normal and the temperature was two degrees warmer all year round (elevated CO_2 climate). The results from the runs are depicted in Figure 3.

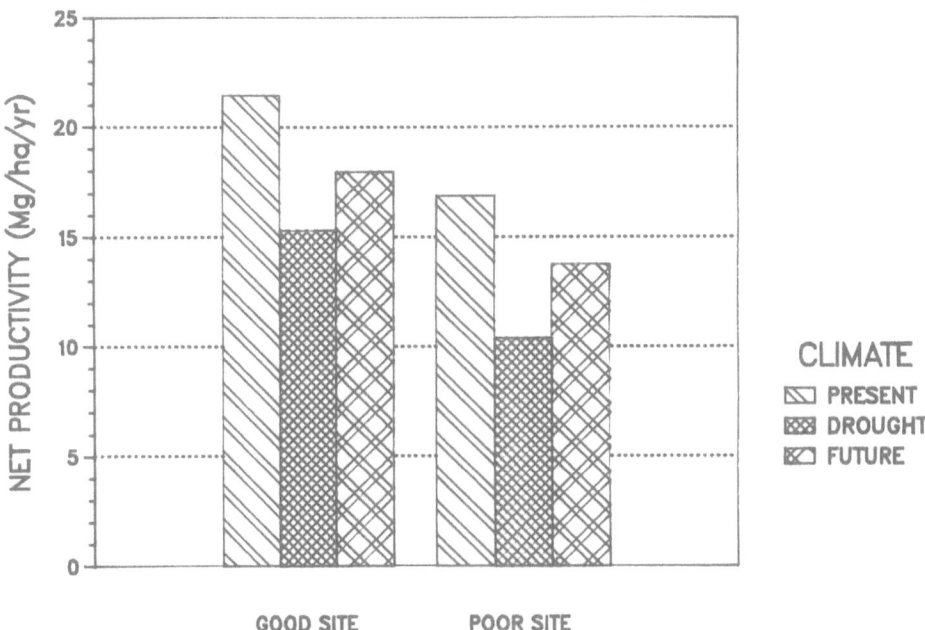

FIGURE 3. The effect of an extended drought and a future climatic shift due to elevated atmospheric CO_2 on net productivity of loblolly pine on a good and poor site in Arkansas

For both sites productivity was reduced more in the drought scenario than the elevated CO_2 scenario. Forests in this region typically experience and survive droughts of this magnitude at least once within a rotation. Results from the model suggest that loblolly pine stands in this region could survive a climatic shift albeit with a 20% decrease in productivity. It is important to note, however, that the model does not consider forest regeneration or seasonal phenology, both of which might be severly impacted by a permanent climatic shift.

The model predicts that forests growing on the poor site would suffer greater relative reduction in productivity due to a drought than forests growing on the good site (a 39% loss versus a 29% loss). A shift in climate would affect both sites with a similar 17-19% loss in productivity.

Strategic planning for the future requires consideration of eventualities such as these. By using the model in such a manner as in our second example, one can make a better educated guess as to the outcome of a hitherto never experienced stress. Certainly the real outcome is still unknown, but if decisions need to be made a model may be helpful.

8.6 SUMMARY

The other chapters in this volume have dealt with individual factors that exert stress on forest ecosystems. However, very seldom does a single stress in isolation impact a forest. A forest responds to the combined effect of many simultaneous stresses. The examples presented in this chapter demonstrate how modeling may be used to examine in a holistic manner the impact of multiple stresses on forest productivity. Modeling can contribute towards developing an integrated understanding of the effect of stress and aid in predicting the outcome of simultaneous stresses. The key to effective modeling lies in defining the objectives of the model and recognizing the limits of the model. We believe stress management and stress management research has and will continue to benefit from the judicious use of modeling.

REFERENCES

1. Aber, J.D., D.B. Botkin, and J.M. Melillo. 1979. Predicting the effects of different harvesting regimes on productivity and yield in northern hardwoods. Can. J. For. Res. 9:10-14.
2. Aber, J.D., D.B. Botkin, and J.M. Millilo. 1978. Predicting the effects of different harvesting regimes on forest floor dynamics in northern hardwoods. Can. J. For. Sci. 8:306-315.
3. Aber, J.D., J.M. Mellilo, and C.A. Federer. 1982. Predicting the effects of rotation length, harvest intensity and fertilization on fiber yield from northern hardwood forests in New England. For. Sci. 28:31-45.
4. Aber, J.D. and J.M. Mellilo. 1981. FORTNITE: A computer model of organic matter and nitrogen dynamics in forest ecosystems. University of Wisconsin Research Bulletin R3130, 31 pp.
5. Agren, G.I. 1985. Theory for growth of plants derived from the nitrogen productivity concept. Physiol. Plant 64:17-28.
6. Agren, G.I. 1983. Nitrogen productivity of some conifers Can. J. For. Res. 13:494-500.
7. Colbert, J.J., W.S. Overton and C. White. 1979. Documentation of the Douglas-fir tussock moth outbreak population model. USDA Forest Service General Technical Report PNW-89. 85 pp.
8. deWitt, C.T. and J. Goudriaan. 1974. Simulation of ecological processes. Puduc Publishers, Wageningen. 159 pp.
9. Doley, D. 1982. Photosynthetic productivity of forest canopies in relation to solar radiation and nitrogen cycling. Aust. J. For. Res. 12:245-261.
10. Emmingham, W.H. and R.H. Waring. 1976. An index of photosynthesis for comparing forest sites in western Oregon. Can. J. For. Res. 7:165-174.
11. Graham, R.L., P. Farnum, R. Timmis and G. Ritchie. 1985. Using modeling as a tool to increase forest productivity and value. R. Ballard, P. Farnum, G.A. Ritchie, and J.K. Winjum (eds). IN: Forest potentials - Productivity and Value. Weyerhaeuser Science Symposium. 4:101-130.
12. Jarvis, P.G. 1981. Plant water relations in models of tree growth. Studia Forestalia Suecica. 160:51-60.
13. Kaufmann, M.R. 1984. A canopy model (RW-CWU) for determining transpiration of subalpine forests I. Model development. Can. J. For. Res. 14:218-226.
14. Kaufmann, M.R. 1984. A canopy model (RW-CWU) for determining transpiration of subalpine forest II. Consumptive water use in two watersheds. Can. J. For. Res. 14:227-232.

15. Kimmins, J.P., K.A. Scouller and M.C. Feller. 1979. FORCYTE-A computer simulation approach to evaluating the effect of whole tree harvesting on nutrient budgets and future forest productivity. S.P. Gessel, R.M. Kenady and W.A. Atkinson (eds). Proceedings of Forest Fertilization Conference. Sept. 25-27, 1979, Alderbrook, Washington. University of Washington College of Forest Resources. Institute of Forest Resources Contribution N40. pg. 266-273.

16. Landsburg, J.J. and R. McMurtie. 1984. Water use by isolated trees. Agric. Water. Manag. 8:223-242.

17. McMurtie, R. and L. Wolf. 1983. A model of competition between trees and grass for radiation, water and nutrients Ann. Bot. 52:449-458.

18. McMurtie, R. and L. Wolf. 1983. Above-ground and below-ground growth of forest stands: A carbon budget model. Ann. Bot. 52:437-448.

19. Mohren, G.M.J., C.P. van Gerwan and C.J.T. Spitters. 1984 Simulation of primary production in even-aged stands of Douglas-fir. For. Ecol. Manag. 8:27-49.

20. Norman, J.M. and P.J. Jarvis. 1985. Photosynthesis in Sitka Spruce (Picea sitchensis (Bong.) Carr.). V. Radiation penetration theory and a test case. J. Appl. Ecol. 12:839-878.

21. Oker-Blom, P. and S. Kellomaki. 1983. Effect of grouping of foliage on within-stand and within-crown light regime: A comparison of random and grouping canopy models. Agric. Meteor. 28:143-155.

22. Oker-Blom, P. and S. Kellomaki. 1982. Theoretical computations on the role of crown shape in the absorption of light by forest trees. Math. Biosci. 59:291-311.

23. Oker-Blom, P. and S. Kellomaki. 1982. Effect of angular distribution of foliage on light absorption and photosynthesis in the plant canopy: Theoretical computation. Agric. Meteor. 26:105-116.

24. Oker-Blom, P. and S. Kellomaki. 1981. Light regime and photosynthesis in the canopy of a Scots pine stand during a prolonged period. Agric. Meteor. 24:185-199.

25. Running, S.W. 1984. Documentation and preliminary validation of H2OTRANS and DAYTRANS, two models for predicting transpiration and water stress in western coniferous forests. USDA Forest Service Research Paper RM-252. 45 pp.

26. Sollins, P., R.A. Goldstein, J.B. Mankin, C.E. Murphy and G.L. Swartzman. 1980. Analysis of forest growth and water balance using computer ecosystem models. D.E. Reichle (ed). IN: Dynamic Properties of Forest Ecosystems. Cambridge University Press, Cambridge. pg. 537-565.

ACKNOWLEDGEMENTS.   We would like to thank Norma Lupkes and
Tina Allen for their patience and perserverence in typing this
manuscript.

9. FIRE: ITS EFFECTS ON GROWTH AND PHYSIOLOGICAL
    PROCESSES IN CONIFER FORESTS

J. L. CHAMBERS, P. M. DOUGHERTY, AND T. C. HENNESSEY
Associate Professor, School of Forestry, Wildlife, and
Fisheries, Louisiana Agricultural Experiment Station,
Louisiana State University, Baton Rouge, LA 70803, Assistant
Professor, School of Forest Resources, University of Georgia,
Athens, GA 30602; and Associate Professor, Department of
Forestry, Oklahoma State University, Stillwater, OK 74078

ABSTRACT

Fire damage impacts tree growth and survival through
direct and indirect effects. Direct effects often involve
cell or tissue death as a result of high internal
temperatures. Crown consumption and crown scorch are the
most visible symptoms of fire injury to pines and other
conifers, but root and cambial damage have also been reported.
Growth and survival appear to be closely correlated to the
degree of crown scorch or crown consumption and tree vigor at
the time of the fire. However, both height and diameter
growth of young pines have been shown to decline significantly
after a fire even though little or no visible damage occurred.
The response of older pines to crown scorch or crown
consumption is similar to that of conifers to pruning and
defoliation.

Indirect effects of fire damage to tree growth and
mortality may occur through fire-caused changes in soil
microorganisms, nutrient cycling or increased attractiveness
and susceptability of the trees to insects and disease. The
use of fire in forest management will probably continue to
increase in the future. As the use of fire increases, risks
and tradeoffs will have to be made. These risks and tradeoffs
are discussed in detail. More research is needed to minimize
the risks and to make the most efficient tradeoffs. Several
suggestions for future research of fire effects on tree growth
are outlined in this paper.

9.1 INTRODUCTION

Fire has played an important role in the ecology of many forest-types for centuries. The role fire has played is quite varied and includes its influence on tree growth; species composition; degree of competition for water, minerals, and light; and seed dispersal, etc. Other roles fire has played include cultural influences, (e.g. disease control, site preparation, etc.) and provision of good wildlife habitat. In the past several decades forest managers have increasingly used prescribed fire as a management tool. Today, there are probably very few knowledgeable forest managers in the Southeast who would not recommend the use of prescribed fire during some stage of stand management. Recently, a chronological bibliography of fire ecology and fire use in the southern pine forest was published (9). With this bibliography Crow gives a brief, but well documented historical journey of fire use. We will, therefore, not cover the development of using prescribed fire in modern southern pine management nor the techniques of prescribed burning.

The purpose of this paper is to review the kinds of information available on the effects of fire damage on the growth of conifers, to make some inferences about the physiological processes which are affected by fire, and then to make some suggestions about the types of fire research needed.

9.2 THEORETICAL MODES OF ACTION
9.2.1 Direct effects

Heat from forest fires can affect trees by heating the soil and air around the tree and its various organs or by heating the tree directly. Transfer of heat occurs by radiation, convection, and conduction. All of these sources of heat transfer are important in determining the impact of fire on tree growth.

Temperature of the flames themselves may approach 1100°C, while air temperatures in convection columns may rise to nearly 540°C (16). Wind and the type of fire have dramatic

effects on the temperature and duration of heat. Headfires
produce hotter conditions for short periods, while slow moving
backfires produce longer periods of sustained heat. In
addition, many factors such as wind speed, fuel moisture, fuel
type, fuel quantity, air temperature, plant temperature,
individual plant characteristics, relative humidity, stand
density, degree of crown closure, and topography also alter
the impacts which fires have on tree growth and survival.

9.2.1.1 Critical or lethal cell temperature: Heat-
caused cell or tissue death is probably the clearest reason
for tree death. Hare (16), in a review of the pertinent
literature on heat effects on plants, reported on research
which indicated that lethal temperature varied from
approximately 54°C to 66°C with the time taken to cause cell
death or seedling death varying from several minutes at the
lower temperature to almost instantaneous death at the
higher temperature. However, Ursic (35) found that roots of
loblolly pine seedlings were more sensitive to temperature
with death occurring between 48°C and 54°C depending on length
of exposure. Nelson (30) found that needle death could be
produced by immersing needles of southern pine species in
water between 54°C and 64°C; again exposure time required for
mortality varied from several minutes at the lower temperature
to a few seconds at the higher temperature.

The insulative properties of tree bark are important in
keeping mortality from fire to a minimum in many trees.
Vines (37) indicated that bark thickness is the main factor
influencing the insulating properties of bark and that
moisture content and structure have little effect on the
insulating properties for most species. Hare (17) tested
several species and found significant differences among
species in the time it took to reach lethal temperature even
though bark thickness was the same. However, he did not
control initial cambial temperatures which varied by almost
7°C. These variations in initial cambial temperature may
explain the differences in the time it took the cambium of
various species to reach the critical temperature.

9.2.1.2 <u>Root damage</u>: Probably one of the least
apparent direct effects of fire on tree growth and survival is
root damage from heat of the fire. Roots have a somewhat
lower critical temperature than stems (35). According to a
review of the literature on fire effects on microorganisms by
Ahlgren (2) soil surface temperatures can often approach 540°C
to 815°C during a fire. Ahlgren (1) reported soil
temperatures of 300°C at a position between the humus and the
mineral soil of a jack pine (<u>Pinus</u> <u>banksiana</u> Lamb.) stand
during a prescribed burn. Hare (16) reported that soil
temperatures at 3.2 to 6.4 mm below the soil surface in a
longleaf pine stand reached 135°C. While reductions in
temperature from the soil surface downward are dramatic, the
impact of elevated soil temperatures may cause death of roots
in the humus and close to the soil surface. These roots are
generally the fine feeder roots. For trees in an otherwise
healthy condition and during favorable environmental
conditions, these roots should be regenerated quickly.
However, for trees in a less vigorous condition or under less
favorable climatic and edaphic conditions, root damage could
be an important factor related to growth loss and mortality.

9.2.1.3 <u>Reduction in leaf area</u>: One of the more
apparent and immediate effects of heat from fires is crown
scorch. Crown scorch is caused by heat kill of the needles
and results in browning of the needles usually within a few
days to a few weeks after the fire. Crown consumption,
actual consumption of needles by the fire, may not be as
apparent if it is slight and restricted to the lower crown.
Substantial reductions in leaf area and therefore in
photosynthetic surface by crown scorch or crown consumption
may reduce, at least temporarily, the trees ability to provide
needed photosynthate for height and diameter growth. Patterns
of transpiration and water use efficiency are necessarily
altered by such crown reduction. Substantial reductions in
leaf area could result in imbalances in the ratio of
photosynthesis to respiration and result in reduced growth and
increased mortality. These trees with a low photosynthesis to

respiration ratio at the time of the fire would most likely experience high mortality. These trees are, also likely to be in the suppressed or intermediate crown classes (3, 14, 27, 36).

9.2.2 Indirect effects

9.2.2.1 Impacts on soil microorganisms: Fire may have indirect effects on tree growth or mortality through its effects on 1) soil microorganisms, 2) impacts on nutrient cycling, 3) changes in susceptibility or attractiveness to insects and disease damage (See Section 9.4.3). Changes in soil fauna and flora following fires have been reported in numerous studies which have been well reviewed by Ahlgren (2). Soil organisms are affected differently based on the type of organism and conditions during and after the fire. Commonly, many soil organisms are killed in the surface soil or litter by fire. Subsequently, relatively rapid increases of these soil organisms are common. Some changes in species have been reported mainly as a result of chemical changes in the soil. Chemical changes (e.g. changes in pH, destructive distillation, etc.) are a result of the fire or may result from changes in surface soil moisture (2). Timing of the fire, fire intensity, soil moisture, and nutrient conditions before and after the fire, as well as the timing of rainfall in relation to the fire, all produce highly variable effects which make it difficult to generalize about the direct effects of fire on soil microorganisms. Also, difficulty in identifying certain organisms by species makes determining the relative benefits and costs of fire on soil microorganisms uncertain.

9.2.2.2 Impacts on organic matter and nutrient cycling: 'While it is difficult to generalize about the effects of fire on the soil microbial population, some evidence is available regarding the impact of fire on the organic matter substrate. Hosking (19) showed humic acids, which made up about 35% of the organic carbon in organic matter, were lost at fire-caused temperatures below 100°C. At temperatures between 100°C and 200°C nondestructive distillation of volatile organic substances occurred and at temperatures between 200°C and 300°C approximately 85% of the substances were destroyed by destructive distillation. DeBano et al. (11) related their findings concerning organic matter to intense, moderate, and light burn in Chaparral conditions with dry soil. They found that intense fires destroyed all organic matter on the surface and most organic matter down to a depth of 2.5 cm in the mineral soil. Moderate burns where surface temperatures reached 432°C destroyed most of the litter. Low intensity fires removed about 85% of the litter on the surface, but only the humic acid components were altered at 2.5 cm depths in the mineral soil.

The loss of organic matter can be greatly reduced by timing the burning, so duff moisture content is moderately high. Martin's (25) guidelines indicate that most prescribed burns should be done when forest floor moisture is moderate to high. This reduces the amount of forest floor removed. Several studies in the southern U.S. have documented the benefits of reducing the fire's impact on the organic matter component of soils (6,28,32). Burning, when forest floor moisture content is moderate, generally leads to increased nutrient concentrations in the upper portion of the mineral soil. In general, prescribed burning on productive forest lands does not have a major impact on growth potential. However, on marginal forest soils with thin layers of organic matter which are not readily incorporated into the mineral soil, the impacts of burning on site productivity could be detrimental.

9.3 OBSERVED GROWTH IMPACTS

Crown scorch and bole char have often been used to categorize the effects of fire on growth and to serve as a diagnostic tool for future work. Most researchers have shown that suppressed trees, small trees, and trees which have been severely scorched suffer the greatest mortality after a fire (3, 8, 14, 16, 26, 34, 36). Analysis of growth in relation to crown scorch has indicated that growth losses, particularly in diameter, are directly related to the degree of crown scorch except at very low scorch levels (20, 26, 36). A study of crown scorch and growth responses in a 20 year old loblolly pine (Pinus taeda L.) plantation after a prescribed burn in southern Louisiana (36) showed diameter growth impacts were directly correlated to the levels of crown scorch. In the first year following the fire, trees with less than one-third of the crown scorched had significantly better diameter growth than trees with unscorched crowns in the same stand. Increased growth at low scorch levels was attributed to elimination of non-productive, "free loading", portions of the crown (36). Trees with from one-third to two-thirds of their crowns scorched had diameter growth approximately equal to unscorched trees, while, trees with crown scorch which covered from two-thirds of the crown to all but the tips of upper crown branches had a 66% reduction in diameter growth. Trees with complete crown scorch had almost no growth (0.02 cm in diameter). After two growing seasons (Villarrubia and Chambers, unpublished data), diameter growth increased substantially in those trees with less than one-third of the crown scorched, stayed the same in those trees with crown scorch ranging from one-third to two-thirds, and remained low in those trees which had experienced greater than two-thirds crown scorch.

A seven year old loblolly pine plantation in southeastern Oklahoma received varying degrees of crown scorch from a prescribed burn in the spring of 1981 (Dougherty 1985, personal communication). Growth data collected from this stand over a period of 4.5 years revealed a loss of 0.73 m (2.4 ft) in height growth and 2.0 cm (0.8 inches) in diameter growth for trees with 30 to 40% crown scorch. Trees with 85 to 99% crown scorch in the same stand were 1.71 m (5.6 ft) shorter and 5.3 cm (2.1 inches) smaller in diameter than trees from adjacent unburned areas of the plantation, 4.5 years after the fire. Measurements immediately after the fire indicated no significant differences in height or diameter at the time of the burn among trees receiving varying degrees of crown scorch.

McCulley (26) indicated that young slash pine (Pinus elliottii Engelm.) trees under 7.6 cm (3 in.) in diameter suffered diameter growth losses even when no crown scorch occurred. He also reported height growth losses occurring in trees up to 17.8 cm in diameter even though no crown scorch occurred. Clason (7) reported that in a 3 year old loblolly pine plantation even scorching of only the lowest branches by a backfire caused a first year growth loss of 8 to 12% in height and a 13% in diameter. In general, the younger the tree the greater the impact of a given level of crown scorch on growth. Several investigators have, however, reported some increase in diameter growth at very low levels of scorch (20, 36).

In still other cases (38), diameter growth after burning has been almost equal to pre-burn levels except under very high fire intensity conditions which have produced high bark char and heavy crown scorch. In Waldrop and Van Lear's study (38) mortality of the intermediate and codominant trees ranged from 20 to 30% when complete crown scorch occurred. A difference of about 1.3 cm in diameter growth was reported between control and high intensity burned plots; however, these differences were not significant.

Growth recovery in stands with varying degrees of crown scorch have generally been indicated to occur within 2 to 3 years. Although long term data is lacking, some research has reported mortality 3 to 4 years after the fire . Recommendations for prescribed burning have ranged from caution and strict adherence to certain weather conditions, to avoidance in young stands, to burning anytime when needed in stands with trees greater than a specified size.

9.4 UNDERLYING CAUSES FOR GROWTH IMPACT

Comparisons of fire damage with damage caused by other factors may provide insight into the effects of fire damage on tree growth. Research areas which may provide information include studies of artificial pruning, studies of artificial or insect caused defoliation, and studies of the effects of lightning damage on a tree's attractiveness to insects and disease. In addition, studies of the relative rates of net photosynthesis in various portions of tree crowns may be useful in assessing where crown damage from fire is likely to have the greatest impact on growth.

9.4.1 Pruning and within crown variation in photosynthesis

Stiell (33) pruned whorls of branches from red pine (Pinus resinosa Ait.) and found that total volume growth was not always proportional to the amount of foliage remaining. Upper branch whorls seemed to be more photosynthetically efficient per unit weight and thus contributed more to volume growth than lower branch whorls. A study by Woodman (40) confirmed this by showing that photosynthesis in a codominant Douglas-fir (Pseudotsuga menziessii (Mirb.) Franco) was more efficient near the top of the crown than in the lower-most branches. Woodman found that from the 18 whorls of branches in the study tree, the uppermost branch whorl produced at a level of 71% of the maximum found within the crown; the seventh branch whorl from the top produced the maximum level of photosynthesis, whorl 12 only 50% of maximum, and whorl 14 just 20% of maximum. Thus, branches within the lower 35% of the crown were producing photosynthate at levels well below 50% of the maximum attained in the upper portion of the crown. In pruning studies by Young and Kramer (43) 14-year-old loblolly pine was pruned by removing approximately 0, 30, and 60% of the existing crown. Cumulative differences in diameter growth were noted within a few weeks after the beginning of the growing season. The greatest reduction in diameter growth occurred at dbh and the reduction in growth became smaller upward through the crown.

Bennett (4) performed similar experiments with 5- and 11-year old slash pine. He found that a 50 percent or greater reduction in crown size caused a significant reduction in diameter growth. Diameter growth reductions were substantial even in the second year but recovery began in the third and fourth year. This is similar to the reported recovery responses pattern observed in trees with fire scorched crowns.

The influence of pruning on tree growth in loblolly pine was also studied by Labyak and Schumacher (22). They followed the effects of pruning of 15-year-old loblolly pine over a period of 15 years. Based on growth of trees which were pruned to leave live crown ratios of 10 to 90% in 10% intervals, they were able to describe resulting growth patterns. Branches nearest the top of the tree contributed most to periodic annual volume growth. Substantial amounts of the lower crown could therefore be removed without greatly reducing volume growth. Branches within the lower 25 percent of the crown with few branchlets were shown not to contribute to main stem growth.

9.4.2 Defoliation and needle contribution

Evenden (13) studied the effects of pine butterfly defoliation on ponderosa pine (Pinus ponderosa Laws.). Mortality was common among defoliated trees as were growth losses and failure to produce annual rings. Of 100 study trees, 84 had greater than 75% defoliation and 31% of those died. Roughly half died from defoliation, while the other were attacked by bark beetles. The period of time without basal growth ranged from 1 to 11 years. Some trees even showed recovery to near pre-defoliation growth rates and then died. O'Neil (31) manually defoliated 13-year-old jack pine in August. He found in 64% of the cases the terminal buds failed to develop if the current years foliage was removed. The greater the proportion of needles removed the greater was the reduction in growth. The greatest reduction in height growth occurred when all current-year needles were removed.

Larson (24) isolated the growth contribution of needles of different ages on red pine. He found that shoot elongation was primarily a function of stored food reserves but that elongation was enhanced by the presence of photosynthesizing needles of various ages. He noted that earlywood development followed elongation of current year needles and that as they matured and elongation ceased, latewood formation began. Trees which are fire damaged in the spring and have severe crown scorch may lose lower limbs and older needles as well as some current year needles. Therefore, production of earlywood may be substantially reduced in the lower bole of the tree.

Continuous needle elongation occurred in Larson's study (24) when only current year needles were exposed to light. Wyant, Laven, and Omi (42) found that the length of developing needles on fire-scorched branches of ponderosa pine was significantly greater than the length of needles developing on unscorched trees.

Timing of defoliation may significantly alter the type of growth changes which occur. Kulman (21) in a review of the literature on the effects of insect defoliation on tree growth and mortality indicated that timing of defoliation completely altered the effects of defoliation. Timing affected needle age which affected hormone levels, photosynthetic efficiency, etc. Timing of fire is also likely to alter the effects of defoliation. Growth loses, no change in growth, differences in the ratio of earlywood to latewood cells and growth pattern shifts in later years are all possible depending on the time and extent of defoliaton.

Webb (39) indicated that growth losses and mortality are heavily dependent on the starch content of twigs at the time of defoliation. Webb pointed out that a tree's ability to put on new crown growth after defoliation was a more sensitive indicator of recovery than bole-wood growth loss. In his study, twig starch content at bud burst was correlated to needle and shoot regrowth after bud burst. Webb also pointed out that the number of needles per shoot was not influenced by defoliation, but the number of shoots was affected since the number of aborted buds per tree was directly affected by defoliation. Therefore, crown biomass recovery was

influenced. These findings are similar to the related effects
on bud abortions found by Evenden (13). Since starch content
of twigs varies with season, species, and environmental
conditions, fire caused defoliation effects on growth will
also likely differ accordingly. The effects of fire caused
defoliation may differ from insect defoliation, as different
portions of the crown are affected.

9.4.3 Lightning effects

Several biochemical changes in trees may occur as the
result of fire damage, crown defoliation, and lightning.
These changes affect the relative vigor of damaged trees and
appear to increase the attraction of insects and disease to
fire damaged trees. Such trees often show an increase in
attractiveness to insect and disease attack.

Blanche, Hodges, and Nebeker (5) studied the effects of
a lightning strike on bark beetle susceptibility indicators in
loblolly pine. After a lightning strike no oleoresin
production occurred for three days. After 11 days, oleoresin
was only 1/10 the normal level. Blanche, Hodges, and Nebeker
pointed out that resin flow was a primary line of defense
against bark beetle attack. They found alpha-pinene, a
primary attractant to the beetles, increased in concentration
by 71% after the lightning strike and myrcene, a precursor to
Ips beetle (Ips spp.) pheromones, increased by 335%. In
contrast they found limonene, which is toxic to bark beetles,
decreased as did the level of beta-pinene, a mountain pine
beetle repellent.

Similar chemical changes may be possible in fire
damaged trees. Hodges (1985, personal communication)
indicated that he had measured substantial declines in
oleoresin pressures following fire damage. Harper (18)
studied the effects of fire on the yields of gum in a naval
stores operation in Georgia. Fire defoliation, simulating
heat defoliation by crown scorch, was applied to 35-year-old
longleaf pines (Pinus palustris Mill.). Crown defoliation of
one-third, two-thirds, and full scorch were evaluated. Gum

yields were reduced from 11 to 32% in the year following the defoliation. Decreasing yield followed increasing defoliation level. Second year losses were not as great as first year losses.

Gerry (15) compared oleorisin production from 40-year-old turpentined longleaf pine which were either severely crown scorched or unscorched by fire. By the beginning of the second growing season the production of oleoresin had been reduced by 50% in the scorched trees. Wood from scorched trees showed retarded development of resin canals, tracheids, and phloem. Effects were still apparent three years after the fire and incomplete annual ring formation occurred in the fire scorched trees.

9.5 RISKS AND TRADEOFFS

Prescribed burning undoubtedly has a place in the management of southern pines as well as many other species. However, from the numerous reports of fire damage even under so called "controlled or prescribed conditions", it becomes apparent that some risks will be taken or tradeoffs will be made if fire is to be used as a management tool. But, how serious will the risks be in various situations and will the tradeoffs be worthwhile? How can the risks be minimized?

9.5.1 Early burning versus early growth

9.5.1.1 Precommercial thinning: Several investigators have suggested or advocated the use of fire as a means of precommercial thinning in young, overly-dense stands (8, 27, 29). The need for precommercial thinning to enhance growth of remaining trees during the rapid growth portion of the rotation has long been justified from a biological standpoint. However, mechanical precommercial thinning has little or no chance of implementation, especially in today's economy. Alternative means of thinning which are low in cost, such as prescribed burning, are, therefore, looked at more seriously. While use of prescribed burning to achieve a precommercial thinning objective may seem like a wise investment, the need for caution and for more information on the effects of fire on tree growth have been expressed (10, 41).

9.5.1.2 <u>Possible growth effects</u>: Growth losses have been reported not only for trees experiencing substantial degrees of crown scorch and crown consumption but also for young trees in burned stands which have shown no crown scorch or other apparent evidence of fire damage. Since most growth losses from prescribed burning have been indicated to last only for a period of a few (2 to 3) years many managers may be willing to give up a little growth early to hopefully reap improved growth in later years. However, as with pruning and defoliation studies, stands with moderate to severe crown scorch and young stands may experience other growth related changes yet to be studied in any detail. Changes in earlywood-latewood ratios, resin production, nutrient allocation patterns, and other processes may also occur. These changes may offer opportunity, as well as problems, and deserve more detailed study.

9.5.2 <u>Fuel reduction burns</u>

Burning for the purpose of fuel reduction has long played an important role in forestry practices in the southeastern U.S.A., where the incidence of human caused fires is high. The failure to use caution, particularly in young stands or older stands which have become too dense and where the fuel has been allowed to accumulate, can be counter-productive. Several studies have indicated that crown scorch may substantially reduce the benefits of prescribed fire as a fuel reduction tool.

de Rhonde (12) investigated the effects of crown scorch on needle fall in a slash pine stand. Results of his study indicated that initial loading of needles was heavy in scorched areas of the stands with an almost 900 percent greater accumulation than normal during the first 103 days following the fire. When averaged across all scorch levels, two years of accumulated needle fall on the scorched areas averaged 140% of that in unscorched areas. In the study area this difference amounted to 2382 Kg/ha of additional needle fuel. Waldrop and Van Lear (38) also reported significant differences in needle fall between areas with scorched and areas with unscorched tree crowns. Eight weeks after the burn, areas with crown scorch had from 1345 to 2466 kg/ha more needle accumulation than areas without scorch.

9.5.3 Changes in nutrient cycling

Numerous studies of the effects of prescribed fires on the nutrient content and availability of forest soils have indicated that under controlled conditions where fires remain cool, little if any significant changes in nutrient availability of the sites occur. When changes in nutrient availability have been reported they have been shown to return to normal within a short period of time. Landsberg and others (23) have found, however, that a prescribed burn with resulting crown scorch on nitrogen-poor ponderosa pine sites led to a reduction in needle biomass and total crown nitrogen content. Even though live crown ratio returned to normal (pre-fire levels) in a few years, needle biomass and total crown nitrogen content remained lower than before the fire. Total foliar N contents were reduced by 14% in moderate fuel reduction burns and 33% in heavy fuel reduction burns and were well correlated to reductions in duff layer thickness. Likewise periodic annual growth was reduced (height by 8 and 18%, volume by 23%). Further study of the impacts of prescribed burning (proper and improper) on the nutrient cycling process of nutrient poor sites is needed.

9.6 RESEARCH NEEDS

Fire has many potential beneficial uses in forestry, but also has the potential for a good deal of harm. The effects of fire on tree growth, yield, and survival are complex and the results of research to date are conflicting. Much more research on fire effects on tree growth is warranted. The use of fire for the sole purpose of benefiting

tree growth will certainly require a significant increase in
fire research. 'We would like to suggest some areas of fire
research we feel are requisite to more efficient and safe use
of prescribed burning in stand management. Our list is not a
complete list, but a partial list of avenues of research which
we feel will be important to understanding the effects of fire
on tree growth and survival.

Many pine sites in the southeastern U.S.A. have limited
nitrogen levels even for tree growth. Landsberg and others
(23) have shown the need for a closer look at the effects of
fire and duff reduction on all aspects of the nutrient cycle
and on biomass accumulation. This research should involve
nutrient allocation or partitioning patterns and the
research should perhaps segregate nutrient sufficient sites
from nutrient poor sites for analysis. Studies which
compare fertilized plots and unfertilized plots might help
answer questions concerning nutrient site interactions and
help separate some environmental effects of fire from the
effects on fire of tree growth and tree growth patterns.

Longterm studies (5 to 10 years) on the impacts of fire
and of fire damage on tree responses are needed if we are to
make wise and frequent use of fire in day to day management
practices. Studies of crown redevelopment patterns after
fire and the changes in radial growth along the tree bole
and within the crown should provide useful information about
the potential impacts of fire and fire damage. These
studies should be coordinated to provide information on
changes in tree form, volume growth, earlywood-latewood
ratios, branch retention patterns, and resulting wood quality
or yield. Research is also needed relative to the impacts of
fire damage on phloem and cambium production.

Perhaps the most beneficial area for more research
would come from research on fire effects in young pine
stands. This is the point at which heavy fuel, competition,
and stand density problems deal their most serious blow to
stand growth and yield. In addition, the ever decreasing

length of rotation merits serious new research into the use
of fire in young stands which are often most susceptible to
fire damage.

REFERENCES
1. Ahlgren, C.E. 1970. Some effects of prescribed burning
    on jack pine reproduction in northeastern Minnesota.
    Minn. Agr. Expt. Stat. Misc. Rep. 94.
2. Ahlgren, I.F. 1974. The effect of fire on soil
    organisms. p. 47-72. In T.T. Kozlowski and C.E. Ahlgren
    (ed) Fire and Ecosystems Academic Press. N.Y. 542 p.
3. Allen, P.H. 1960. Scorch and mortality after a summer
    burn in loblolly pine. USDA-FS Southeastern For. Expt.
    Sta. Res. Note 144.
4. Bennett, F.A. 1955. The effect of pruning on the height
    and diameter growth of planted slash pine. J. For.
    53:636-638.
5. Blanche, C.A., J.D. Hodges and T.E. Nebeker. 1985.
    Changes in bark beetle suceptibility indicators in a
    lightning struck loblolly pine. Can. J. For. Res.
    15(2)397-399.
6. Brender, E.V. and R.W. Cooper. 1968. Prescribed burning
    in Georgia's Piedmont loblolly pine stands. J. For.
    66:31-36.
7. Clason, T.R. 1985. Prescribed burning to improve
    timber-pastures. Louisiana Agric. 29(1) 20-21.
8. Cooper, R.W. and A.T. Altobellis. 1969. Fire kill in
    young loblolly pine. USDA-FS Fire Control Notes 30(4)
    14-15.
9. Crow, A.B. 1982. Fire ecology and fire use in the pine
    forest of the South. A chronological bibliography.
    School of Forestry and Wildlife Management. Louisiana
    State University, Baton Rouge 131 p.
10. Crow, A.B. and C.L. Shilling. 1980. Use of prescribed
    burning to enhance Southern pine timber production. S.J.
    Appl. For. 4(1)15-18.
11. DeBano, L.F, P.H. Dunn and C.E. Conrad. 1977. Fire's
    effect on physical and chemical properties of chaparral
    soils. USDA-FS General Report WO-3-65-74.
12. de Rhonde, C. 1983. Controlled burning in pine stands
    in the Cape: The influence of crown scorch on tree
    growth and litterfall. South African For. J., Vol.
    123:39-41.
13. Evenden, J.C. 1940. Effects of defoliation by the pine
    butterfly upon ponderosa pine. J. For. 38:949-955.
14. Ferguson, E.R. 1955. Fire-scorched trees-will they live
    5or die? p. 102-113. In . Crow, A.B. (ed.) Modern
    Forest fire management in the South. Proc. 4th Ann. For.
    Symp. School of Forestry, Louisiana State Univ. April
    6-7, 1955. Baton Rouge, LA.
15. Gerry, Eloise. 1931. Oleoresin production from longleaf
    pine defoliated by fire. J. Agric. Res. 43:827-836.

16. Hare, R.C. 1961. Heat effects on living plants. USDA-FS South. For. Exp. Sta. Occasional Paper 183. 32 p.

17. Hare, R.C. 1965. Contribution of bark to fire resistance of southern trees. J. For. 63: 248-251.

18. Harper, V.L. 1944. Effects of fire on gum yields on longleaf and slash pine. USDA Circ. 710, 42 p.

19. Hosking, J.S. 1938. The ignition at low temperatures of the organic matter in soils. J. Agric. Sci. 28(3): 393-400.

20. Johansen, R.W. 1975. Prescribed burning may enhance growth of young slash pine. J. of Forestry 73(3) 148-149.

21. Kulman, H.M. 1971. Effects of insect defoliation on growth and mortality of trees. Ann. Rev. Entomol. 16:289-324.

22. Labyak, L.F. and F.X. Schumacher. 1954. The contribution of its branches to the main-stem growth of loblolly pine. J. For. 52:333-337.

23. Landsberg, J.D., P.H. Cochran, M.M. Finck, and R.E. Martin. 1984. Foliar nitrogen content and tree growth after prescribed fire in ponderosa pine. USDA-FS Pacific Northwest For. and Range Exp. Sta. Res. Note PNW-412. 15 p.

24. Larson, P.R. 1964. Contribution of different-aged needles to growth and wood formation of young red pines. For. Sci. 10:224-238.

25. Martin, R.E. 1981. Prescribed burning techniques to maintain or improve soil productivity pp. 66-70. Hubbs, S.D., Helgerson, O.T., eds. Reforestation of skeletal soils: Proceedings of a workshop; Held Nov. 17-19, 1981. Medford, OR. For. Research Laboratory, Oregon State Univ., Corvallis, OR.

26. McCulley, R.D. 1950. Management of natural slash pine stands in the flatwoods of South Georgia and North Florida. USDA Circ. No. 845. 19p.

27. McNab, W.H. 1977. An overcrowded loblolly pine stand thinned with fire. S.J. Appl. For. 1:24-26.

28. Moehring, D.M., C.X. Grano, and J.R. Bassett. 1966. Properties of forested loess soils after repeated prescribed burns. South. For. Exp. Stat., USDA-FS Res. Note. SO-40., 4 p.

29. Morris, W.G. and E.L. Mowat. 1958. Some effects of thinning a ponderosa pine thicket with a prescribed fire. J. of For. 56:203-209.

30. Nelson, R.M. 1952. Observations on heat tolerance of southern pine needles. USDA-FS Southeastern For. Exp. Sta. Station Paper 14, 6 p.

31. O'Neil, L.C. 1962. Some effects of artificial defoliation on the growth of Jack pine (Pinus banksiana LAMB.). Can. J. Bot. 40:273-280.

32. Romancier, R.M. 1960. Reduction of fuel accumulations with fire. Southeast For. Exp. Stn. USDA-FS Res. Note SE 143, 2 p.

33. Steill, W.M. 1969. Stem growth reaction in young red pine to the removal of single branch whorls. Can. J. Bot. 47:1251-1256.
34. Storey, T.G. and E.P. Merkel. 1960. Mortality in a longleaf-slash pine stand following a winter wildfire. J. of Forestry 58(3) 206-210.
35. Ursic, S.J. 1961. Lethal Root temperature of 1-0 loblolly pine seedlings. USDA-FS Tree Planters Notes No. 47, p. 25-29.
36. Villarrubia, C.R. and J.L. Chambers. 1978. Fire: Its effects on growth and survival of loblolly pine, Pinus taeda L. La. Acad. Sci. 41:85-93.
37. Vines, R.G. 1968. Heat transfer through bark, and the resistance of trees to fire. Aust. J. Bot. 16:499-514.
38. Waldrop, T.A. and D.H. VanLear. 1984. Effect of crown scorch on survival and growth of young loblolly pine. S.J. Appl. For. 8(1)35-40.
39. Webb, W.L. 1981. Relation of starch content to conifer mortality and growth loss after defoliation by the Douglas-fir Tussock moth. For. Sci. 27:224-232.
40. Woodman, J.N. 1971. Variation of net photosynthesis within the crown of a large forest-grown conifer. Photosynthetica 5(1)50-54.
41. Wooldridge, D.D. and H. Weaver. 1965. Some effects of thinning a ponderosa pine thicket with a prescribed fire, II. J. of For. 63:92-95.
42. Wyant, J.G. and R.D. Laven and P.N. Omi. 1983. Fire effects on shoot growth characteristics of ponderosa pine in Colorado. Can. J. For. Res. 13:620-625.
43. Young, H.E. and P.J. Kramer. 1952. The effect of pruning on loblolly pine. J. For. 50:474-479.

10. HERBICIDE STRESS: USE OF BIOTECHNOLOGY TO CONFER HERBICIDE
RESISTANCE TO SELECTED WOODY PLANTS

N. D. NELSON AND B. E. HAISSIG

Program Leader and Project Leader, Biotechnology Research
Program, USDA Forest Service, North Central Forest Experiment
Station, Forestry Sciences Laboratory, Rhinelander, Wisconsin

ABSTRACT

The control of weed competition with herbicides represents
a major opportunity for increasing forest yields and decreasing
the costs of production. However, many herbicides damage the
crop trees unless precisely applied. As a consequence, suboptimum weed control is often accepted in current silvicultural practice. Herbicide selectivity for weeds versus crop
can be achieved in three ways; avoidance, tolerance, and
resistance. Broad spectrum herbicide selectivity in tree
plantations is currently limited to avoidance and species-level
tolerance. The new biotechnologies provide a means of impart-
ing herbicide tolerance and resistance to specific tree genotypes, thereby allowing important improvements in weed control.
The availability of herbicide-resistant hardwoods (angiospermous trees) would allow large-scale plantations of these
trees in developed countries. The primary mode of action of
an herbicide is a key to understanding the stress effects of
that chemical on the forest stand and to evaluating it as a
candidate for a biotechnology program. Some of the most
promising ways of achieving herbicide tolerance or resistance
in forest trees are through detoxification (metabolism) of the
herbicide, overproduction of the target enzyme of the herbicide, and alteration of the target protein of the herbicide.
Potentially genes for these characteristics could be incorporated in trees through somaclonal selection and recombinant DNA
technologies. Based on several scientific, economic, and
environmental criteria, glyphosate, the sulfonylureas, and
the imidazolinones are good candidate herbicides for which to

develop resistance in forest trees. Research in this area, initiated in 1984, is described.

10.1 INTRODUCTION

The control of weed competition represents a major opportunity for increasing forest yields (39). Herbicides have been a cost-effective way to decrease losses in agriculture due to weeds (63). Interest has been growing in the use of chemical herbicides in the forestry-sector as well, despite the importance of mechanical methods and prescribed fire as weed control technologies (53). The advantages of herbicides in specific instances include lower cost, usability on a wider range of sites (41), and more effective control of weed re-sprouting (54). Herbicides may be the only practical alternative for herbaceous weed control in southern pine plantations (54). It is likely that herbicides will be increasingly important in worldwide forestry.

Unfortunately, in spite of their general effectiveness, many herbicides can damage the forest tree crop as well as the weeds. This fact limits the flexibility and effectiveness of herbicide use in specific forest management situations and even precludes the planting of some, potentially valuable tree species. With current technology, foresters have important but limited capabilities to ensure the selectivity of herbicides for weeds versus the crop trees. Conventional tree breeding could be used to increase herbicide selectivity. However, this application of breeding is impractical for trees because of the probable low frequency of herbicide-resistance genes in natural tree populations and the long sexual generation times of most tree species (55, 57). The new biotechnologies (including somaclonal selection, somatic hybridization, direct DNA insertion, recombinant DNA) provide a potentially practical means of making trees resistant to effective herbicides. The existence of herbicide-resistant trees could permit great improvements in weed control with resultant large increases in productivity and substantial reductions in costs. The first significant impact of biotechnology in agronomy will

probably be in the herbicide resistance area (16, 35). The
same is likely in forestry (57).

10.2 PRODUCTIVITY AND EFFICIENCY INCREASES THROUGH WEED CONTROL AND HERBICIDE RESISTANCE

Herbicides are used in silviculture to control weed competition in site preparation, plantation establishment, release
(weed control after plantation establishment), and timber
stand improvement. (The term "weed" in this paper refers to
any higher plant competing with the tree crop for moisture,
nutrients, and light, whether herbaceous or woody.) Chemical
weed control has been shown to increase the height, diameter,
and volume growth rates of both conifer and hardwood stands
from 10 up to 1000 percent as compared to untreated controls
(31, 34, 39, 43, 53, 54, 58, 61, 70). An additional benefit
of such increased growth is a shortening of rotation length
with attendant increased profitability (42, 59). The beneficial effects of herbicidal weed control on stand survival are
often even more important than effects on growth alone -- often
meaning the difference between plantation failure and success
(10, 33, 59, 70). The current generally low level of competition control in forest stands, combined with the large beneficial effects of weed control on tree survival and growth, makes
chemical weed control one of the most effective stress management options for forest stands and one of the most promising
yield enhancement opportunities in forestry.

The availability of herbicide-resistant planting stock
for important forest tree species would allow foresters to
gain substantially more benefits from weed control. Resistance
to a specific broad spectrum herbicide would permit that herbicide to be broadcast applied during plantation establishment
or release, and would provide virtually complete weed control
with complete safety for the tree crop. With current technology, applications of broad spectrum herbicides, while the tree
stand is growing, usually are limited to directed spraying.
Directed spraying prevents contact between the crop trees and
the herbicide. Such directed applications vary in effective-

ness but are always suboptimum, because weeds immediately
adjacent to trees cannot be safely treated. Another problem
with directed applications of broad spectrum herbicides in
tree plantations is damage to the crop trees due to accidental
drift, movement in surface water, and leaching of the herbicide
into the root zone of the tree crop.

The suboptimality of directed application of herbicides and
the sensitivity of many important hardwood species to herbicides have severely limited the practice of artificial reforestation with hardwoods (angiospermous trees) in the economically developed countries. Because of these limits, large
hardwood plantation schemes have been located in the developing
countries where hand weeding is economically feasible because
of low labor costs. The availability of herbicide-resistant
hardwoods could contribute strongly to a restructuring of
plantation forestry and the creation of new large-scale hardwood plantation crops in the developed countries. Although
foresters typically have more flexibility in using herbicides
in conifer plantations than in hardwood stands, the existence
of herbicide resistance genes in conifer planting stock would
add substantially to the efficacy of weed control in conifer
plantations as well.

Another major benefit of herbicide resistance would be
reductions in the cost of plantation establishment and management. Multiple entries of the plantation are typically
required to effectively control weeds with current technology.
Herbicide resistance would reduce the required number of weed
control entries by allowing a much greater degree of weed
control from each application of the chemical. The optimum
case for such cost reduction may be herbicide resistance involving soil active, broad spectrum herbicides of low mammali-
an toxicity with some persistence of immobile activity on the
site. Reductions in the required incidences of herbicide
usage may also have environmental benefits.

10.3 HERBICIDE MODE OF ACTION

The primary mode of action of an herbicide is a key to

understanding the stress effects of that chemical on the
forest stand and to evaluating it as a candidate for a biotechnology program. The other key is knowledge of herbicide
metabolism and detoxification mechanisms, which will be discussed under "Opportunities for Managing Herbicide Stress"
in this paper.

There are many ways of characterizing the physiology and
biochemistry of herbicide mode of action. The application of
molecular biology to the study of herbicide action has greatly
increased knowledge about the sites of action of herbicides
over the last few years (35). The reader should consult one
of the recent comprehensive books on herbicide action for a
detailed treatment (eg. 18, 24). For the present purposes,
herbicides can be classified by the level of cellular organi-
zation required for activity and by the site(s) of primary
activity within the plant cell.

There are essentially two classes of herbicides: (a)
herbicides that require that a specific specialized tissue be
present for activity and (b) herbicides that affect a cellular
function or functions present in every plant cell. The pre-
dominant version of the former are the photosynthetic inhibitors. Important examples of photosynthetic inhibitor herbicides in current silvicultural practice are the triazines,
substituted ureas, and bipyridiliums. Herbicides that impinge
a universal cellular function do not require specialized plant
tissues for activity and include a wide variety of chemicals.
Important forestry examples of general cellular function herbicides include 2,4-dichlorophenoxyacetic acid (2,4-D), picloram,
glyphosate, and amitrole. General cellular function herbicides
may be more immediately amenable than the photosynthetic
inhibitors to inclusion in biotechnology programs. In vitro
selection for resistance to the photosynthetic inhibitor herbicides may require the development of photoautotrophic or
photomixotrophic in vitro cultures, the inclusion of an early
selection step involving differentiated shoots or whole plants,
or both. In contrast, biotechnology experiments involving the
general cellular function herbicides may require only

undifferentiated cells for the initial selection procedures.

The primary site of activity of an herbicide within the plant cell can be characterized as four alternatives: (a) inhibit the activity of a single protein; (b) disrupt membrane function; (c) affect multiple sites; (d) unknown.

The first alternative may be of greater interest than the others in biotechnology and genetic engineering. Most of the conclusive information on single protein targets for specific herbicides has been accumulated in the last five years, due largely to the revolutionary advances in molecular biology and their application to plant science (35). Herbicides in this category include the s-triazines, glyphosate, sulfonyl-ureas, and imidazolinones (35). All these chemicals have secondary and lower order biochemical and physiological effects, in addition to their primary or direct effect (19). The relative simplicity of single gene-single protein systems is a major advantage in the study and manipulation of such herbicides. All herbicides proven to have a single protein site of primary activity block biochemical pathways not found in mammals (19) although this may be conditioned by the toxi-cological requirements for the commercial development of modern herbicides. These pathways are photosynthesis (s-tria-zines) and essential amino acid biosynthesis (glyphosate, sulfonylureas, imidazolinones).

10.4 HERBICIDE STRESS EFFECTS ON FOREST STANDS

Herbicide damage to tree crops ranges from slight reduc-tion in growth rate to total mortality, the exact effect being specific to the chemical, the amount of the chemical applied, the method of application, site factors, meteorological con-ditions during and following application, and the genotype, morphology, age, and phenological state of the tree crop. These effects cannot be generalized. In some cases, reduced growth, visible injury, and mortality in the tree stand are easily discerned (eg. 1, 32, 69). However, herbicidal stress on tree crops is probably often overridden and hidden by the positive effects of weed control on tree growth.

The conifers are generally more tolerant of herbicides than are hardwoods, although there are many exceptions. The tolerance of conifers to the phenoxy herbicides, including 2,4-D and 2,4,5-trichlorophenoxyacetic acid (2,4,5-T) which have been used extensively for killing hardwood competition in conifer stands (17, 25, 27, 59), is a classic and commercially important example. The s-triazine, simazine, can injure or kill many hardwoods but can be safely sprayed on conifer foliage in any season (72). Norway spruce (<u>Picea</u> <u>abies</u>) apparently achieves its tolerance to simazine through its ability to metabolize this chemical (47). Hexazinone is registered for release of several conifer species from hardwoods and other competition (20, 27). Recent data suggest that loblolly pine (<u>Pinus</u> <u>taeda</u>) may be tolerant of hexazinone because of its ability to metabolically detoxify the chemical (48). Some conifers have exhibited seasonal tolerance for glyphosate (17, 52, 53, 60, 71) although the factors involved in this seasonality are only partially understood (52, 60). Glyphosate normally severely injures or kills hardwoods when applied to them at rates recommended for weed control (53, 71), with resistance only being achieved during dormancy (eg. 2).

10.5 OPPORTUNITIES FOR MANAGING HERBICIDE STRESS

10.5.1 Herbicide selectivity

Herbicide selectivity is the ability to control weeds with a chemical without harming the crop. Herbicide selectivity can be achieved in three ways: (a) avoidance, (b) tolerance, (c) resistance.

10.5.1.1 <u>Avoidance</u>: Herbicide avoidance is the prevention of effective contact between the chemical and the crop or responsive tissues of the crop. Avoidance is an important method of preventing damage to tree stands in current silvicultural technology. It is based on separating the chemical and the tree crop or responsive tissues of the tree crop in space or in time. Spatial separation is achieved through directed application of the chemical, so that weeds are contacted by the herbicide but susceptible tissues of the tree stand are not. Temporal separation is achieved by applying the herbicide at times of the year when the tissues of the tree crop are not responsive to the chemical while weed populations are. Such application times are sometimes referred to as "windows," most commonly including the late growing season, early quiescence, early dormancy, and late dormancy periods of the tree crop. The specification of such safe windows probably depends on the genotype, age, and phenological and general physiological state of the tree crop, site factors, meteorological conditions, and the weed populations present. Little research has been done on the physiology of temporal herbicide avoidance for forest tree crops. There are recent publications on this subject for glyphosate (52, 60). The large size and perennial nature of trees provide many opportunities for herbicide avoidance with tree crops. Consequently, avoidance may be even more important for tree crops than for agronomic crops.

10.5.1.2 Tolerance and resistance: We define herbicide tolerance as the innate ability of a genotype(s) of a crop to suffer much less injury than the norm for the crop species from an herbicide that is in effective contact with the crop, at a dosage that controls competing weeds. The incorporation of herbicide tolerance in forest trees could be a significant commercial goal, of potentially even more importance than in the case of agronomic crops. In agronomy, total herbicide resistance may be necessary to qualify as acceptable as a primary value-added feature of a new crop variety. Herbicide-tolerant tree planting stock could be particularly efficacious because of the latitude for using herbicide avoidance techniques in forestry. Combining herbicide tolerance and avoidance in a holistic vegetation management system for a particular type of plantation silviculture may result in weed control as cost effective as that possible with herbicide-resistant tree planting stock.

We define herbicide resistance as complete immunity of a

genotype(s) of a crop to a specific herbicide or herbicides in
effective contact with the crop, at dosages at or above those
levels that control competing weeds. Herbicide resistance is
a much rarer trait than herbicide tolerance in natural, un-
treated plant populations (46).

Herbicide tolerance and resistance can be achieved in at
least six ways: (a) reduced uptake, (b) reduced translocation,
(c) intracellular sequestration, (d) detoxification (metabolism), (e) target enzyme overproduction, (f) target protein
alteration.

There are many examples of herbicide tolerance of a crop
species, and herbicide susceptibility of a weed species, being
the result of differential absorption or translocation of the
chemical (37). However, interaction may occur among
differential absorption, degradation (metabolism) near the
site of absorption, and translocation in producing such relative tolerance and susceptibility. Thus, the etiology of such
effects is often complex (37). Phenological and morphological
variation strongly conditions differential absorption in many
cases. The primary causes of true differential translocation
of herbicides are not well understood (37). It seems likely
that most instances of herbicide tolerance due to reduced
uptake or translocation, separate from differential degradation, are crop traits controlled by many genes. The complexity, lack of understanding of causal mechanisms, and probable
multigenic control of reduced uptake and translocation as
primary effectors of herbicide tolerance make this area
difficult to approach using biotechnological techniques. In
addition, uptake and translocation of herbicides are differentiated functions of whole plants or organs. It might be
difficult or impossible to induce the expression of these
functions in cell or tissue culture in a manner that relates
meaningfully to whole plant characteristics. Thus, selection
and testing for herbicide tolerance based on reduced uptake
or translocation of the herbicide must be done at the whole
plant level, with attendant large increases in costs.

Intracellular sequestration of herbicides away from their

site of action has been proposed as a mechanism of resistance
(26). Evidence was presented suggesting that paraquat resistance in a biotype of Conyza bonariensis was due to exclusion
of paraquat from its site of action in the chloroplast by a
rapid compartmentation mechanism (26). This mechanism appears
to be a promising approach for achieving resistance in plants,
but little information is yet available on the frequency of
occurrence, genetic control, and biochemistry of this phenomenon.

Metabolic detoxification or degradation of herbicides as
a primary mechanism of plant tolerance is a well-documented
phenomenon with promise for biotechnological manipulation.
The most important basic biochemical reactions in higher
plants that result in herbicide detoxification are oxidation,
reduction, hydrolysis, and conjugation(65). Other less common
or less well documented detoxification mechanisms include
acylation, alkylation, cyclization, and ring cleavage (36).
The genetic engineering (direct DNA insertion, recombinant DNA)
strategy for effecting plant tolerance through such reactions
is to identify and isolate genes encoding enzymes involved in
herbicide detoxification. The gene for such an enzyme would
then be transferred to the plant, where the production of the
enzyme would inactivate the herbicide, preventing the herbicide
from acting at its target site in the cell (16).

The two major sources of herbicide detoxification genes
are (a) pre-existing tolerant plant species, and (b) microorganisms isolated from soils and water where high concentrations of herbicides have been present over relatively long
time periods (16).

There are many examples of herbicide-tolerant higher
plant species with naturally occurring specific herbicide
detoxification systems (16). One of the most broadly occurring
systems is the conjugation of herbicides with the tripeptide,
glutathione, catalyzed by the glutathione-S-transferases
(EC 2.5.1.18) (65). Glutathione conjugation is also a major
mechanism for detoxification of xenobiotics in animals (11).
Maize (Zea mays) is a well-known example of a plant exhibiting

herbicide tolerance because of a large contribution from the natural presence of the glutathione conjugation system. A major factor in the tolerance of maize to triazines is the presence of triazinyl-glutathione-S-transferase (65, 66).

Several microbes that degrade herbicides have been iden- tified and studied (16). For example, a Pseudomonas species has been identified which can metabolize glyphosate (50). Genes encoding the microbial enzymes that catalyze such degradations are candidates for transfer to crop plants through recombinant DNA or other biotechnologies to effect herbicide tolerance or resistance in the crop plant. To our knowledge, no higher plant has yet been transformed in this manner.

The overproduction of enzymes essential for growth has been observed in inhibitor-tolerant cell lines of microbes (35), mammals (3), and higher plants (4, 51). Examples specific to herbicide tolerance include (a) bacteria toler- ant of glyphosate, which overproduce the target enzyme of glyphosate, 5-enolpyruvylshikimate-3-phosphate synthase (EPSP synthase, EC 2.5.1.19) (4, 40, 62); (b) yeast tolerant of the sulfonylurea, sulfometuron methyl, which overproduce the target enzyme of the sulfonylureas, acetolactate synthase (ALS, EC 4.1.3.18) (22); (c) cell suspension cultures of the higher plant, Corydalis sempervirens, which are tolerant of glyphosate and which overproduce EPSP synthase (4); (d) cell suspension cultures of carrot (Daucus carota), tolerant of glyphosate, which also overproduce EPSP synthase (51); and (e) a line of alfalfa (Medicago sativa) tissue culture cells tolerant of the herbicide L-phosphinothricin, a competitive inhibitor of gluta- mine synthase, which overproduce this target enzyme (29). In two of the microbial examples (22, 40, 62) and in the alfalfa cell line (29), the overproduction of the target enzyme of the herbicide resulted at least partly from the presence of multi- ple copies of the gene encoding for the target enzyme; i.e., gene amplification. The only instance of herbicide tolerance due to target enzyme overproduction at the whole plant level of which we are aware is glyphosate-tolerant Petunia plants

which were transformed by recombinant DNA methods by Monsanto
Corporation scientists to overproduce EPSP synthase (verbal
presentations at 1985 WSSA and ISPMB Meetings by R. Fraley).

Most of the research on the biotechnology of herbicide
resistance has been focused on modifying the site of action
of the herbicide (28). This is the "altered target" (16) or
"structural gene mutation" (35) approach. The usual strategy
is to cause a single amino acid change in the protein that is
the primary inhibitory target of the herbicide in the cell,
resulting in a protein that is not affected by the herbicide
if the correct amino acid substitution has been made. The
point change in the target protein is effected by a corre-
sponding single nucleotide base pair change in the gene that
encodes the protein. The single gene nature of such transfor-
mations and the theoretical stability of the resistances so
engendered make this an attractive approach for achieving
herbicide resistance (35).

There are several recent examples of microbes and higher
plants that are resistant to specific herbicides as a result
of alterations in the amino acid composition of the target
protein. Single cell organisms exhibiting this form of resistance include Chlamydomonas, resistant to atrazine (38),
Salmonella and E. coli, resistant to glyphosate (15), and
Salmonella (45) and Saccharomyces (22, 23), resistant to
sulfonylureas. The only herbicides for which there is yet
unequivocal publicly-available proof of resistance from target
protein alteration in higher plants are the triazines and
sulfonylureas. More than 30 species of atrazine-resistant
weeds have been identified which appear to have such resistance (30). In addition, atrazine-resistant canola (Brassica
napus) based on this form of herbicide resistance is now
commercially available (8, 9). Tobacco (Nicotiana tabacum)
plants, resistant to sulfonylureas, have been selected from
tissue culture (14). The basis of the resistance in these
tobacco plants was subsequently proven to be the presence of
a form of the target enzyme of the sulfonylureas, ALS, that is
insensitive to this class of herbicides (13).

Extensive research is in progress to transfer resistance
to sulfonylureas and glyphosate into higher plants by transferring resistant target proteins isolated from microorganisms
via recombinant DNA and possibly other biotechnologies (16,
35). Preliminary reports indicate that tolerance, but not yet
resistance, to glyphosate has been achieved in tobacco at the
whole plant level in this work (5).

The recent applications of molecular biological approaches
have produced a revolutionary increase in our knowledge of
herbicide target proteins. In addition to the identification
of EPSP synthase as a major target enzyme of glyphosate and
ALS as the major target of the sulfonylureas described above,
the target proteins for the imidazolinones and the triazines
have been elucidated. The target enzyme of the imidazolinones
has also been identified as ALS (64), even though the
imidazolinones bear little chemical resemblance to the
sulfonylureas. EPSP synthase and ALS catalyze essential amino
acid biosynthetic reactions and are encoded by nuclear genes
(35). The triazine target is a 32 Kd membrane protein in the
photosystem II complex of the chloroplasts, designated the QB
protein (7, 38). QB protein is encoded by a chloroplast gene,
psbA.

Herbicide selectivity in current silvicultural practice is
based on avoidance and species-level tolerance. These modes
of selectivity provide important but limited methods of herbicide use in forestry. The new biotechnologies for the first
time provide potential means of achieving more powerful forms
of herbicide selectivity based on genotype-specific tolerance
and resistance. The successful transfer of herbicide tolerance
and resistance traits to elite forest tree germplasm is likely
to result in much more flexible and effective use of herbicides
in forest practice and, therefore, to more efficient and economic forestry.

10.5.2 Biotechnology of herbicide tolerance and resistance

The new biotechnologies of possible use in transferring
herbicide tolerance and resistance to crop plants include

somaclonal selection, somatic hybridization (protoplast fusion), direct DNA insertion (microinjection, liposome technology, others), and recombinant DNA. At the current stage of development of these technologies, the most promising for application to forest trees are somaclonal selection and recombinant DNA.

Somaclonal selection is the selection of specific somaclonal variation. Somaclonal variation is the mendelian and non-mendelian genetic variation detected in plants derived from any form of cell or tissue culture (21, 44). Many examples already exist of higher plant cell lines, whole plants from tissue culture, and progeny of whole plants from tissue culture that exhibit stable herbicide tolerance and resistance as a result of selection in vitro (6, 12, 14, 35, 49, 67). To our knowledge, the only such example for a woody plant is nucellar callus culture lines of orange (Citrus sinensis) stably tolerant of 2,4-D (68).

Recombinant DNA work on herbicide resistance is emphasizing the transformation vector, Agrobacterium. The prime examples are glyphosate and sulfonylurea resistance. The status of this work that is public information was described in the "Herbicide Selectivity" section. Comai and Stalker (16) and Hardy and Giaquinta (35) provide good reviews of work in progress in this area.

Research on the use of somaclonal selection and recombinant DNA for transferring herbicide tolerance and resistance to forest tree species was just initiated in 1984. This promising work will be described later in this paper.

The ideal herbicide for which to develop resistance in forest trees should have the following characteristics:

. broad spectrum activity (lethal to a wide variety of higher plants)

. lethal to both herbaceous and woody weeds

. active pre- and post-emergence

- low toxicity to mammals and other animal life forms and low potential for adverse environmental effects in general

- inhibition of a cellular function(s) present in every cell

- known primary site of cellular activity

- resistance encoded by a nuclear gene

- resistance monogenic

- resistance based on altered protein target

- chemical insensitivity to *in vitro* culture media

- *active* concentration readily quantifiable in *in vitro* culture

- lethal to undifferentiated cell and tissue cultures at concentrations of same order of magnitude as effective concentrations used in field applications

- lethal to microbial species that are amenable to gene isolation and cloning, at physiological concentrations of the herbicide

- resistance already achieved in one or more nonforest higher plant species

- approval from major regulatory agencies for forestry uses or reasonable probability that approval will be sought.

Three other characteristics -- soil activity, environmental persistence, and cost -- will be evaluated differently as properties of an herbicide selected for crop resistance incorporation, depending on the observer. Soil activity and persistence of herbicidal activity in the field may often be desirable characteristics of a forestry herbicide from the viewpoint of the silviculturist because of the common need for season-long and multi-year weed control in forest plantations. However, soil activity and persistence may sometimes be real or perceived negative attributes of an herbicide to an environmentalist.

Cost of the herbicide may also be perceived in opposite ways in relation to desirability of candidate herbicides for resistance incorporation, depending on the position of the evaluator in the market dynamics of the chemical. For example, the forest manager will desire cheap herbicides, other factors being equal. Low cost of an herbicide generally equates with chemicals that are "off patent." Such herbicides may also be chosen for the development of resistance in crop plants by chemical companies that continue to sell large quantities of such herbicides after they are off patent, with the objective of increasing their market share for the chemical. Alternatively, herbicides that are under patent and command a high unit price (per unit of land area treated) will be the most desirable candidates for inclusion in herbicide resistance research programs for the chemical companies holding the patents. The forest manager may also favor the higher priced herbicide as a candidate for herbicide resistance development if no cheaper herbicide offers good alternative weed control.

Based on all considerations, at least three herbicides stand out as good candidates for resistance development in forest tree species: glyphosate (Roundup); the sulfonylurea, sulfometuron methyl (Oust); and the imidazolinone, 2-(4-iso-propyl-4-methyl-5-oxo-2-imidazolin-2-yl) nicotinic acid (Arsenal). Roundup, Oust, and Arsenal are trademarks of Monsanto, DuPont, and American Cyanamid Corporations, respectively. A checklist of the characteristics of these herbicides that relate to their suitability as candidates for herbicide resistance research with forest trees is given in Table 1.

TABLE 1. Candidate herbicides for tolerance/resistance development in forest trees through biotechnology.

	glyphosate	sulfometuron methyl	2-(4-isopropyl-4-methyl-5-oxo-2 imidazolin-2-yl) nicotinic acid
Broad spectrum	Yes	Yes	Yes
Herbaceous & woody weeds	Yes	No, mainly herbaceous	Yes
Pre- & post-emergence	Post only	Yes	Yes
Mammalian toxicity & general environmental effects	Low toxicity, environmentally safe	Low toxicity environmentally safe	Low toxicity, environmentally safe
Inhibit general cellular function	Yes	Yes	Yes
Primary site of action known	Yes, EPSP synthase	Yes, ALS	ALS
Resistance encoded by nuclear gene	Yes	Yes	Yes
Resistance mono-genic	Yes	Yes	Yes
Altered target protein resistance	Yes, EPSP synthase	Yes, ALS	Yes, ALS
Deactivated by incorporation in cell culture media	Yes, at least partially	No	No
Active concentration easily quantifiable in vitro	No	Yes	Yes
Kills undifferentiated cells at physiological concentrations	?, may be genotype specific in vitro	Yes	?
Kills microbial species amenable to genetic engineering, at physiological concentrations	Yes	Yes	?
Resistance achieved in non-forest higher plant species	Tolerance only	Yes	Yes
Registered for forestry uses	Yes	Yes	Experimental use permit only in U.S.

Soil activity	No	Yes	Yes
Persistence	No	Yes	Yes
Low cost	No, under patent	No, under patent	No, under patent

Other broad spectrum herbicides may be used for herbicide resistance research with forest trees in the future, but at the present state of biotechnology, they are viewed as slightly less promising because of individual features. The triazines (simazine, atrazine, hexazinone) are promising in all respects except that one of the two major known forms of resistance is encoded by a chloroplast gene. No predictable transformation systems for chloroplast genes have yet been developed. In addition, these herbicides require the presence of photosynthetic tissue for activity at physiological concentrations. These characteristics of the triazines may make recombinant DNA approaches for target protein alteration impossible and somaclonal selection approaches difficult with current technology for these chemicals. The transfer of triazine metabolism genes to trees may be a possibility, however. Other possible herbicides, albeit with one or more drawbacks for herbicide resistance work based on current knowledge, include amitrole, 2,4-D, picloram, triclopyr, and paraquat. Other forestry herbicides that we cannot yet evaluate as candidates for resistance incorporation include asulam and fosamine. New developments in biotechnology could rapidly alter the plausibility of working with a given herbicide. In addition, new herbicides from the chemical industry may show greater promise in the future than any of the herbicides discussed above.

10.5.3 Current biotechnology research

In 1984, the North Central Forest Experiment Station Biotechnology Program of the USDA Forest Service initiated the first research on the biotechnology of herbicide resistance in forest trees (Table 2) (55, 56, 57). This research includes both somaclonal selection and recombinant DNA approaches; the recombinant DNA project is in cooperation with Calgene, Inc.

and the University of Wisconsin (Table 2).

TABLE 2. Current research on the biotechnology of herbicide
tolerance/resistance in forest trees.

Study	Organization	Herbicide	Species	Biotechnology emphasis
Genetic modulation of somaclonal variation in herbicide tolerance in natural populations of poplar	North Central Forest Exp. Sta.(NC) Biotechnology Program (USDA Forest Service)	sulfonyl-ureas	Populus	somaclonal selection
Herbicide resistant elite hybrid poplar clones through somatic cellular variation	NC Biotechnology Program	glyphosate, sulfonyl-ureas	Populus	somaclonal selection
Conferring glyphosate resistance on selected Populus through genetic transformation with aroA gene	NC Biotechnology Program; Calgene, Inc.; University of Wisconsin	glyphosate	Populus	recombinant DNA

Other research organizations have shown substantial interest in the Forest Service's entry into this research area. As a result, it is likely that other research projects on the biotechnology of herbicide resistance in forest trees have been recently initiated or soon will be. All known research in this area is using Populus as the experimental subject, because this tree genus is amenable to manipulation and regenerability in vitro and is internationally important for fiber and biomass production. It is likely that such research will be expanded to other important forest tree species in the future, including the conifers. The success of such work with conifers will probably depend on improvements in the ability to regenerate whole plants of conifers from cell and tissue culture systems.

REFERENCES

1. Akinyemiju, O. A., and D. I. Dickmann. 1982. Variation among 21 Populus clones in tolerance to simazine and diuron. Canadian Journal of Forest Research 12: 708-712.

2. Akinyemiju, O. A., J. G. Isebrands, N. D. Nelson, and
    D. I. Dickmann. 1982. Use of glyphosate in the estab-
    lishment of Populus in short rotation intensive culture.
    Pages 161-169. In: Proc. N. Amer. Poplar Council Meet-
    ing. Rhinelander, Wisconsin. Kansas State University,
    Division of Extension, Manhattan, Kansas.
3. Alt, F. W., R. E. Kellems, J. R. Bertino, and R. T.
    Schimke. 1978. Selective multiplication of dihydrofolate
    reductase genes in methotrexate-resistant variants of
    cultured murine cells. Journal of Biol. Chem. 253: 1357-
    1370.
4. Amrhein, N., D. Johanning, J. Schab, and A. Schulz. 1983.
    Biochemical basis for glyphosate-tolerance in a bacterium
    and a plant tissue culture. FEBS Letters 157: 191-196.
5. Anonymous. 1985. Inserted gene makes tobacco herbicide
    resistant. Genetic Technology News 5(3): 1.
6. Anonymous. 1985. Selection of herbicide-tolerant corn
    plants. Agricell Report 4: 5.
7. Arntzen, C. J., and J. H. Duesing. 1983. Chloroplast-
    encoded herbicide resistance. Pages 273-294. In:
    Advances in Gene Technology: Molecular Genetics of
    Plants and Animals (F. Ahmand, K. Downey, J. Schultz, and
    R. W. Voellmy, eds.). Academic Press, New York.
8. Beversdorf, W. D., J. Weiss-Lerman, and L. R. Erickson.
    1980. Registration of triazine-resistant Brassica
    campestris germplasm. Crop Sci. 20: 289.
9. Beversdorf, W. D., J. Weiss-Lerman, L. R. Erickson, and
    V. S. Machado. 1980. Transfer of cytoplasmically
    inherited triazine resistance from birds rape to culti-
    vated oil seed rape Brassica campestris and Brassica napus
    cultivar tower. Canadian Journal of Genetics and
    Cytology 22: 167-172.
10. Bey, C. F., J. E. Krajicek, R. D. Williams, and R. E.
    Phares. 1976. Weed control in hardwood plantations.
    Pages 69-84. In: Herbicides in Forestry (W. R. Byrnes
    and H. A. Holt, eds.). Purdue University, Department of
    Forestry and Natural Resource, West Lafayette, Indiana.
11. Boyland, E., and L. F. Chasseaud. 1969. The role of
    glutathione and glutathione S-transferases in mercapturic
    acid biosynthesis. Advances in Enzymology 32: 173-219.
12. Chaleff, R. S. 1981. Genetics of higher plants. Appli-
    cations of cell culture. Cambridge University Press,
    Cambridge, U.K. 184 p.
13. Chaleff, R. S., and C. J. Mauvais. 1984. Acetolactate
    synthase is the site of action of two sulfonylurea
    herbicides in higher plants. Science 224: 1443-1445.
14. Chaleff, R. S., and T. B. Ray. 1984. Herbicide-resist-
    ant mutants from tobacco cell cultures. Science 223:
    1148-1151.
15. Comai, L., L. C. Sen, and D. M. Stalker. 1983. An
    altered aroA gene product conifers resistance to the
    herbicide glyphosate. Science 221: 370-371.
16. Comai, L., and D. M. Stalker. 1984. Impact of genetic
    engineering on crop protection. Crop Protection 3: 399-
    408.

17. Conard, S. G., and W. H. Emmingham. 1983. Herbicides for shrub control on forest sites in northeastern Oregon and northern Idaho. Oregon State University, College of Forestry, Forest Research Laboratory Special Publication 5. 8 p.

18. Duke, S. O. (Ed.). 1985. Weed physiology. Volume II. Herbicide physiology. CRC Press, Boca Raton, Florida. 257 p.

19. Duke, S. O. 1985. Effects of herbicides on nonphotosynthetic biosynthetic processes. Pages 91-112. In: Weed Physiology. Volume II. Herbicide Physiology (S. O. Duke, ed.). CRC Press, Boca Raton, Florida.

20. DuPont, Inc. 1984. Grow more wood in less time with DuPont forestry herbicides. E. I. du Pont de Nemours & Co., Agricultural Chemicals Dept., Wilmington, Delaware.

21. Evans, D. A., W. R. Sharpe, and H. P. Medina-Filho. 1984. Somaclonal and gametoclonal variation. American Journal of Botany 71: 759-774.

22. Falco, S. C., and K. S. Dumas. 1985. Genetic analysis of mutants of Saccharomyces cerevisiae resistant to the herbicide sulfometuron methyl. Genetics 109: 21-35.

23. Falco, S. C., K. S. Dumas, and R. E. McDevitt. 1984. Molecular genetic analysis of sulfonylurea herbicide action and resistance in yeast. Pages 467-478. In: Molecular Form and Function of the Plant Genome. (L. van Vloten-Doting, G. S. P. Groot, and T. C. Hall, eds.). NATO ASI Series A: Life Sciences Volume 83.

24. Fedtke, C. 1982. Biochemistry and physiology of herbicide action. Springer-Verlag, New York. 202 p.

25. Freed, V. H. 1984. Chemical and biological agents in forest pest management. Historical overview. Pages 1-23. In: Chemical and Biological Controls in Forestry (W. Y. Garner and J. Harvey, Jr., eds.). ACS Symposium Series 238. American Chemical Society, Washington, D.C.

26. Fuerst, E. P., H. Y. Nakatani, A. D. Dodge, D. Penner, and C. J. Arntzen. 1985. Paraquat resistance in Conyza. Plant Physiol. 77: 984-989.

27. Gjerstad, D. H. 1981. Chemical weed control in southern forests. Pages 116-120. In: Proc. Weed Control in Forest Management. The 1981 John S. Wright Forestry Conference (H. A. Holt and B. C. Fischer, eds.). Purdue University, West Lafayette, Indiana.

28. Glaser, V. P. 1983. Researchers tackle herbicide resistance. Bio/Technology 1: 826-827.

29. Goodman, H. M., E. Tischer, and G. Donn. 1984. Herbicide resistance in plants: an example of gene amplification. Journal of Cellular Biochem. (S) 8B: 148.

30. Gressel, J., Y. Regev, S. Malkin, and Y. Kleifeld. 1983. Characterization of an s-triazine-resistant biotype of Brachypodium diastachyon. Weed Science 31: 450-456.

31. Guldin, R. W. 1984. Economic returns from spraying to release loblolly pine. Pages 248-254. In: Proc., Biotechnology and Weed Science, Southern Weed Science Society, 37th Annual Meeting. Hot Springs, Arkansas. Weed Science-Society of America, Champaign, Illinois.

32. Hallett, R. D. 1983. Seedling injury by simazine and other triazine herbicides. Environment Canada, Maritimes Forest Research Centre, Frederickton, New Brunswick. Information Report M-X-145. 13 p.

33. Hansen, E., L. Moore, D. Netzer, M. Ostry, H. Phipps, and J. Zavitkovski. 1983. Establishing intensively cultured hybrid poplar plantations for fuel and fiber. U.S.D.A. Forest Service, NC Forest Experiment Station, St. Paul, Minnesota. General Technical Report NC-78. 24 p.

34. Hansen, E., D. Netzer, and W. J. Rietveld. 1984. Weed control for establishing intensively cultured hybrid poplar plantations. U.S.D.A. Forest Service, NC Forest Experiment Station, St. Paul, Minnesota. Research Paper NC-317. 6 p.

35. Hardy, R. W. F., and R. T. Giaquinta. 1984. Molecular biology of herbicides. BioEssays 1: 152-156.

36. Hatzios, K. K., and D. Penner. 1982. Metabolism of herbicides in higher plants. Burgess Publishing Company, Minneapolis, Minnesota. 142 p.

37. Hess, F. D. 1985. Herbicide absorption and translocation and their relationship to plant tolerances and suscepti- bility. Pages 191-214. In: Weed Physiology. Volume II. Herbicide Physiology (S. O. Duke, ed.). CRC Press, Boca Raton, Florida.

38. Hirschberg, J., A. Bleecker, D. J. Kyle, and L. McIntosh. 1984. The molecular basis of triazine-herbicide resist- ance. Z. Naturforsch. 39: 412-420.

39. Holt, H. A., and B. C. Fischer (eds.). 1981. Proc. Weed Control in Forest Management. The 1981 John S. Wright Forestry Conference. Purdue University, West Lafayette, Indiana. 305 p.

40. Jaworski, E. G., T. J. Mozer, S. G. Rogers, and D. C. Tiemeier. 1984. Herbicide target sites, mode of action, and detoxification: chloroacetanilides and glyphosate. Pages 335-349. In: Biosynthesis of the Photosynthetic Apparatus: Molecular Biology, Development and Regulation (P. J. Thornber, A. L. Staehelin, and R. B. Hallick, eds.). Alan R. Liss, New York.

41. Kellison, R. C., and S. Gingrich. 1984. Loblolly pine management and utilization--state of the art. Southern Journal of Applied Forestry 8: 88-96.

42. Kerr, E. 1982. Herbicides offer promise for lower site preparation costs. Forest Farmer 41(10): 13-16.

43. Knowe, S. A., L. R. Nelson, D. H. Gjerstad, B. R. Zutter, G. R. Glover, P. J. Minogue, and J. H. Dukes, Jr. 1985. Four-year growth and development of planted loblolly pine on sites with competition control. Southern Journal of Applied Forestry 9: 11-14.

44. Larkin, P. J., and W. R. Scowcroft. 1981. Somaclonal variation - a novel source of variability from cell culture for plant improvement. Theoret. Appl. Genet. 60: 197-214.

45. LaRossa, R. A., and J. V. Schloss. 1984. The sulfonyl- urea herbicide sulfometuron methyl is an extremely potent and selective inhibitor of acetolactate synthase in

Salmonella typhimurium. Journal of Biological Chemistry 259: 8753-8757.

46. LeBaron, H. M., and J. Gressel. 1982. Herbicide resistance in plants. John Wiley & Sons, New York. 401 p.

47. Lund-Hoie, K. 1969. Uptake, translocation and metabolism of simazine in Norway spruce (Picea abies). Weed Research 9: 142-147.

48. McNeil, W. K., J. F. Stritzke, and E. Basler. 1984. Absorption, translocation, and degradation of tebuthiuron and hexazinone in woody species. Weed Science 32: 739-743.

49. Meredith, C. P., and P. S. Carlson. 1982. Herbicide resistance in plant cell cultures. Pages 275-291. In: Herbicide Resistance in Plants (H. M. LeBaron and J. Gressel, eds.). John Wiley & Sons, New York.

50. Moore, J. T., H. S. Braymer, and A. D. Larson. 1983. Isolation of a Pseudomonas which utilizes the phosphonate herbicide glyphosate. Applied and Environmental Microbiology 46: 316-320.

51. Nafziger, E. D., J. M. Widholm, H. C. Steinrucken, and J. L. Killmer. 1984. Selection and characterization of a carrot cell line tolerant to glyphosate. Plant Physiol. 76: 571-574.

52. Neal, J. C., and W. A. Skroch. 1985. Growth stage effects on glyphosate absorption and transport in ligustrum and juniper. WSSA Abstracts 24: 35.

53. Nelson, L. R., D. H. Gjerstad, and P. J. Minogue. 1984. Use of herbicides for industrial forest vegetation management in the southern United States. Pages 11-23. In: Chemical and Biological Controls in Forestry (W. Y. Garner and J. Harvey, Jr., eds.). ACS Symposium Series 238. American Chemical Society, Washington, D.C.

54. Nelson, L. R., D. H. Gjerstad, and P. J. Minogue. 1984. Herbicides for practical forest management in the South. Forest Farmer 43(5): 19-21.

55. Nelson, N. D. 1985. North Central Forest Experiment Station Biotechnology Program -- application to tree improvement. Pages 14-22. In: Proc. 1985 Southern Forest Tree Improvement Conference. Long Beach, Mississippi.

56. Nelson, N. D., and B. E. Haissig. 1984. Biotechnology in the Forest Service's North Central Forest Experiment Station. Pages 139-154. In: Proc. International Symposium of Recent Advances in Forest Biotechnology. Traverse City, Michigan. Michigan Biotechnology Institute, East Lansing, Michigan.

57. Nelson, N. D., B. E. Haissig, and D. E. Riemenschneider. 1984. Applying the new somaclonal technology to forestry. Pages 27-34. In: Proc. TAPPI 1984 Research and Development Conference. Appleton, Wisconsin. TAPPI, Atlanta, Georgia.

58. Netzer, D. A. 1984. Herbicide trials on European larch in northern Wisconsin. U.S.D.A. Forest Service, NC Forest Experiment Station, St. Paul, Minnesota. Research

214

Note NC-318. 2 p.

59. Newton, M. 1981. Chemical weed control in western forests. Pages 127-138. In: Proc. Weed Control in Forest Management. The 1981 John S. Wright Forestry Conference (H. A. Holt and B. C. Fischer, eds.). Purdue University, West Lafayette, Indiana.

60. Perala, D. A. 1985. Using glyphosate herbicide in converting aspen to conifers. U.S.D.A. Forest Service, NC Forest Experiment Station, St. Paul, Minnesota. Research Paper NC-259. 6 p.

61. Ponder, F., Jr., and R. C. Schlesinger. 1984. Site influences herbicide efficiency and growth of planted hardwoods. Forest Ecology and Management 9: 147-153.

62. Rogers, S. G., L. A. Brand, S. B. Holder, E. S. Sharps, and M. J. Brackin. 1983. Amplification of the aroA gene from Escherichia coli results in tolerance to the herbicide glyphosate. Applied and Environmental Microbiology 46: 37-43.

63. Schaub, J. R. 1985. The economics of agricultural pesticide technology. Pages 15-26. In: Agricultural Chemicals of the Future. Beltsville Symposia in Agricultural Research 8 (J. L. Hilton, ed.). Rowman & Allanheld, Totowa, New Jersey.

64. Shaner, D. L., P. C. Anderson, and M. A. Stidham. 1984. Imidazolinones. Potent inhibitors of acetohydroxyacid synthase. Plant Physiol. 76: 545-546.

65. Shimabukuro, R. H. 1985. Detoxification of herbicides. Pages 215-240. In: Weed Physiology. Volume II. Herbicide Physiology (S. O. Duke, ed.). CRC Press, Boca Raton, Florida.

66. Shimabukuro, R. H., D. S. Frear, H. R. Swanson, and W. C. Walsh. 1971. Glutathione conjugation. An enzymatic basis for atrazine resistance in corn. Plant Physiol. 47: 10-14.

67. Singer, S. R., and C. N. McDaniel. 1985. Selection of glyphosate-tolerant tobacco calli and the expression of this tolerance in regenerated plants. Plant Physiol. 78: 411-416.

68. Spiegel-Roy, P., J. Kochba, and S. Saad. 1983. Selection for tolerance to 2,4-dichlorophenoxyacetic acid in ovular callus of orange (Citrus sinensis). Z. Pflanzenphysiologie 109: 41-48.

69. South, D. B. 1984. Response of loblolly pine and sweetgum seedlings to oxyfluorfen. Auburn University, Southern Forest Nursery Management Cooperative Publication Number 16. 14 p.

70. United States Department of Agriculture. 1984. Effects of competing vegetation on forest trees: a bibliography with abstracts. U.S.D.A. Forest Service. General Technical Report WO-43.

71. Wendel, G. W., and J. N. Kochenderfer. 1984. Aerial release of Norway spruce with Roundup in the central Appalachians. Northern Journal of Applied Forestry 2: 29-32.

72. White, D. P. 1976. Herbicides for weed control in coniferous plantations. Pages 60-68. In: Herbicides in

Forestry (W. R. Byrnes and H. A. Holt, eds.). Purdue University, Department of Forestry and Natural Resources, West Lafayette, Indiana.

ACKNOWLEDGEMENTS

A portion of the herbicide resistance research of the NC Forest Experiment Station Biotechnology Program is supported by the U.S. Department of Energy, Biomass Energy Technology Division, Short Rotation Woody Crops Program under Interagency Agreement No. DE-AI05-800R20763. We gratefully acknowledge the technical reviews of E. Patrick Fuerst and Wesley P. Hackett.

DISCLAIMER

Mention of trade names does not constitute endorsement by the U.S. Department of Agriculture.

PESTICIDE PRECAUTIONARY STATEMENT

This publication reports research involving pesticides. It does not contain recommendations for their use, nor does it imply that the uses discussed here have been registered. All uses of pesticides must be registered by appropriate State and/or Federal agencies before they can be recommended.

CAUTION: Pesticides can be injurious to humans, domestic animals, desirable plants, and fish or other wildlife--if they are not handled or applied properly. Use all pesticides selectively and carefully. Follow recommended practices for the disposal of surplus pesticides and pesticide containers.

11. AIR POLLUTION: SYNTHESIS OF THE ROLE OF MAJOR AIR POLLUTANTS IN
DETERMINING FOREST HEALTH AND PRODUCTIVITY

IVAN J. FERNANDEZ

Assistant Professor of Soil Science and Cooperating Assistant Professor
of Forest Resources, Department of Plant and Soil Sciences, University
of Maine, Orono, Maine   04469

Maine Agricultural Experiment Station Publication 1058, University of
Maine, Orono, Maine

ABSTRACT

Forested ecosystems are exposed to not only a physical climate, but
also a modern chemical climate which has the potential to alter forest
health and productivity. Recently, unexplained forest deterioration
has been identified throughout extensive areas in central Europe, at
high elevation coniferous sites, and apparently in commercial
forestlands in the eastern United States. Deposition of pollutants
from the atmosphere is suspect in playing a role in these forest
deterioration phenomena. To date, no cause and effect relationship has
been conclusively demonstrated between regional versus local
atmospheric deposition and forest deterioration. The major hypotheses
currently thought to deserve the focus of scientific investigation deal
with forest alterations due to (1) gaseous pollutants (primarily O_3),
(2) a fertilization effect (primarily from N), (3) acid deposition
effects on foliage and soils, (4) trace metals (e.g. Pb, Cu, Cd, Ni,
Zn), (5) general stress from air pollutants, and (6) deposition of
organic growth altering substances. Not enough is yet known to suggest
adjustments in forest management decisions which might reduce the
possible effects of atmospheric deposition. However, foresters must
keep informed about the current state-of-knowledge on the air pollution
and forest effects issue as we all will play a role in management and
policy decisions which shape the character of the environment in which
future forests will grow.

11.1 INTRODUCTION

During recent years we have become aware that virtually all
components of the landscape are exposed to not only the physical
climate (i.e. weather), but also a modern chemical climate comprised
of numerous chemical species generated both naturally and through the
activities of man. The concern over "acid rain" has grown over the
last two decades to a point where today this term is often used to more
generally describe "atmospheric deposition," or the deposition of both
wet and dry forms of acidifying and non-acidifying materials from the
atmosphere. The scientific research community studying atmospheric
deposition phenomena has often organized research to study potential
effects on either aquatic resources (e.g. lakes and streams), crops,
forests, or materials (e.g. buildings, statuary, metals). In the late
1970's public attention in the United States was primarily focused on
possible acid rain effects on lakes, with particular concern for
research suggesting that acidic lakes in the Northeast may have
resulted from long-range transported air pollutants. During the early
1980's, reports of deteriorating forests in West Germany and at high
elevation sites in the eastern United States resulted in a dramatic
increase in both public concern (and an emphasis in research) on the
possible role of air pollution in these forest deterioration
phenomena. At present, few questions have answers regarding the
effects of regional air pollution on forested ecosystems. Numerous
studies are underway and planned to address this problem. However, the
task of conclusively identifying cause and effect relationships between
forest health and specific pollutants will be difficult. Forested
ecosystems are complex and typically dominated by tree species which
grow for decades or centuries.

A great deal of literature exists on the atmospheric deposition
effects issue and the reader is advised to consult U.S. EPA (1983a),
CAST (1984), or NSEPB (1982) for documents summarizing information on
this subject. Suggestions for documents dealing specifically with the
forest effects issue include Smith (1981), Bernsten et al. (1984),
Morrison (1984), U.S.F.S. (1985), Schutt and Cowling (1985), Burgess
(1984), Fernandez (1983), Johnson and Siccama (1983), Breece and
Hasbrouck (1984), and TARF (1985). The following discussion is

intended to provide a broad overview, from the author's perspective, of the current situation regarding our understanding of the potential effects of atmospheric deposition on forests.

11.2 AIR POLLUTANTS OF PRIMARY CONCERN

While we often think of the lower atmosphere as being composed of 79% N_2, 21% O_2, and approximately 0.03% CO_2, there are numerous other organic and inorganic chemical species present which may have effects on natural ecosystems. Since public awareness and the political process often drive the direction of scientific research, we seem to sequentially move through periods where attention is focused on selected pollutants within the air pollution and forest effects issue. The evolution of regional forest deterioration phenomena in recent years is resulting in a need to synthesize what we know about the effects on forests of all possible air pollutants. Thus we are beginning to ask questions about the implications of multiple and interactive pollutant exposure for forest health.

Currently the scientific debate on the possible role of air pollution on regional forest health appears to be focused on three categories of air pollutants: (1) acid deposition, (2) photochemical oxidants (primarily O_3), and (3) trace metals (e.g. Pb, Zn, Ni, Cu, Cd). Acid deposition refers to both wet and dry materials deposited from the atmosphere which are acidic or have acidifying properties. Wet acid deposition is primarily due to oxides of sulfur (S) and nitrogen (N) which are further oxidized in the atmosphere to form the strong mineral acids H_2SO_4 and HNO_3. These acids may be transported long distances in the atmosphere and deposited at remote locations in the form of rain, snow, hail, sleet, or cloud and fog moisture. A significant component of HNO_3 may also be deposited in the vapor phase. It is worthy to note that in many high elevation ecosystems, the interception of cloud droplets by the forest canopy can result in large increases of water and chemical inputs in addition to rain and snow. Lovett et al. (1982) estimated that inputs from cloud droplet impaction in a northern Appalachian subalpine balsam fir forest amounted to 46 percent of the bulk precipitation for water, and from 150 to 430 percent of bulk precipitation for the various chemicals.

Dry acid deposition includes the absorption of gaseous nitrogen and sulfur compounds by water, vegetation, or soil, as well as the deposition of particles containing sulfur and nitrogen.

Photochemical oxidants are formed in the atmosphere by sunlight catalyzed reactions between chemicals. Ozone (O_3) is currently considered to be the most important phytotoxic air pollutant in the eastern United States (Chevone, 1985) and is formed in the atmosphere from the reaction between nitrogen oxides and volatile organic compounds (e.g. hydrocarbons) in the presence of sunlight. Ozone and its precursors can be transported long distances in the atmosphere and thus are considered, along with acid deposition, to be among the major air pollutants of concern with regard to regional air pollution effects on forest health. Potentially harmful O_3 near the earth's surface should not be confused with the layer of O_3 in the upper atmosphere which is beneficial, due to its ability to filter ultraviolet light from solar radiation.

Many of the processes of combustion which contribute O_3 precursors, S, and N to the atmosphere also emit a number of trace metals (e.g. Pb, Cu, Cd, Ni, Zn) which can be transported long distances in the atmosphere and deposited as wet and dry forms in forested ecosystems. Some of these metals (e.g. Cu, Zn) are essential plant nutrients and all are typically found in very small quantities in soils and vegetation. Therefore, if the amount of trace metals in soils and vegetation is significantly increased above natural levels, the potential exists for toxic effects on both soil microflora and trees. This group of metals is also often referred to as "heavy metals." These metals have been shown to accumulate in the forest floor and are thought to have the potential to affect forest health either individually, through interactions between metals, or through interactions with other natural or anthropogenic stress factors.

11.3 FOREST DETERIORATION SYMPTOMOLOGY

"Forest declines" have occurred throughout history, often affecting single species and typically disappearing with time, leaving no causative mechanism evident despite scientific investigations of the problem. Recently, reports of deteriorating forest health in central

Europe, southwest Sweden, and in the eastern United States have created a great deal of concern as to the cause or causes of the symptoms, and for the possible role of air pollution in these widespread phenomena. Over 50 percent of West Germany's forestlands are reported to show some symptoms of damage (von Osten, 1985); however, these estimates include damage which may be related to known causes, the anticipated losses of trees common to all forests, as well as damage likely to be due to air pollutants. Some high elevation coniferous forests in the Appalachian corridor of the U.S. are also reported to be declining (Johnson and Siccama, 1983), and indications of commercial forest growth declines have been reported. Forest deterioration in these areas, and possibly in other areas, seems characteristic of stress induced phenomena since no causative mechanisms have yet been defined and several species over a wide geographic area are involved.

Throughout the reports and scientific literature a wide range of symptoms have been associated with these modern forest deterioration phenomena. These symptoms can be organized into three major categories as follows:

(A) Loss of Foliar Biomass

The most readily visible symptom of forest deterioration or decline is the loss of foliar biomass. In the United States this symptom is best illustrated by the "spruce decline" reported for red spruce (Picea rubens Sarg.) at high elevation sites in the northeast. This symptom is characterized by a loss of foliage beginning in the crown and progressing downward and inward over time (Johnson and Siccama, 1983). Red spruce decline has also been documented for some sites in the southern Appalachian Mountains. In West Germany the symptoms include "canopy ghosting" (i.e. a gradual thinning usually of the lower crown resulting in progressive crown transparency) and active casting of green leaves and shoots (Schutt and Cowling, 1985). The species most affected in Europe are white fir (Abies alba Mill.) and Norway spruce (Picea abies Karst.). Scots pine (Pinus sylvestris L.), European beech (Fagus sylvatica L.), and other species have also recently been reported to be involved in this forest disease syndrome the Germans refer to as "Waldsterben."

Fine root biomass also is reduced in diseased individuals. Schutt and Cowling (1985) reported that comparative studies in Europe between healthy and deteriorating old fir, spruce, and beech showed that affected trees had fewer living fine roots, little or no mycorrhizae, less ability to form new feeder roots, and a greater frequency of infection by secondary pathogens.

(B) Altered Physiology or Morphology

We can hypothesize that alterations in tree physiology must occur to bring about the losses in biomass described above, but little research results are available to define the physiological response of trees involved in forest deterioration phenomena. Cowling's (1984) summarization of symptoms of forest deterioration for the U.S. Environmental Protection Agency points to evidence of alterations including foliar chlorosis (i.e. yellowing of foliage), changes in the relative length of long and short shoots, "Stork's nest" crowns in white fir, concentrations of leaves in tufts or clumps at the tips of branches in broadleaved species, changes in leaf morphology, excessive production of adventitious shoots on branches, and excessive production of seeds and cones.

(C) Decreased Annual Increment

A less visible symptom often associated with these forest deterioration phenomena is the decline in the annual increment growth rate of trees. Most of the research documenting this symptom has not looked at total height growth in order to determine the net effects on total biomass production. What is important to note is that many of the trees which have exhibited foliar biomass losses in the past few years show evidence of declining annual increment growth rates beginning one or more decades ago. It is largely due to the visible symptoms of deterioration that we have become aware of the deterioration phenomena. Due to the kinds of measurements made for standard forest inventories, subtle changes in tree growth which may occur in otherwise healthy looking trees may easily go undetected. While no evidence exists to show that widespread growth declines are occurring in the eastern United States, the possibility that subtle reductions may be occurring undetected in growth rates of commercial forests could have significant economic implications.

11.4 GENERAL HYPOTHESES ON FOREST EFFECTS MECHANISMS

In most complex issues of science, we are better at generating hypotheses than we are at testing them. The potential role of air pollution in modern forest deterioration phenomena is no exception, with numerous mechanisms of forest effects having been proposed. Often we find that soil scientists propose soil related mechanisms or physiologists propose physiological mechanisms, as would be expected, since researchers will investigate the areas of science they know best. Scientists often focus their research on particular sites which have specific characteristics, and therefore possibly unique mechanisms of response to air pollution, that may be markedly different from sites studied by other scientists investigating the forest deterioration problem. While no single pollutant has been shown to be the primary causative factor in all declining forests, several air pollutant related hypotheses of forest effects have been proposed which appear to be possible contributing factors in the unexplained deterioration of forests. Table 1 lists general headings for six major air pollutant related mechanisms which this author believes are currently held within the scientific community as the most viable.

Table 1 – Current general hypotheses to describe mechanism of air pollution effects on regional forest health

(1) Gaseous Pollutants (primarily O_3)
(2) Fertilization Effect (primarily N)
(3) Acid Deposition
(4) Trace Metals
(5) General Stress
(6) Growth Altering Substances

All, none, or various combinations of these pollutant stresses may be appropriate for a specific site, and the critical pollutants may differ for different sites. All of these pollutant stresses are likely to be interactive with other natural stresses on forests, particularly with pests and pathogens.

(1) Gaseous Pollutants

On a regional basis (rather than near point sources of emissions) O_3 is considered to be the most damaging gaseous air pollutant in the United States (Chevone, 1985; Skelly, 1980).

Ozone has been documented as a causal factor of the deterioration of mixed coniferous forests in the San Bernadino and San Gabriel Mountains of southern California (Miller et al., 1982) where chronic elevated O_3 levels exist. In the eastern United States, episodic O_3 stress of forests has become apparent with species such as eastern white pine (Pinus strobus L.), shown to be particularly susceptible to this air pollutant. Davis and Wilhour (1976) have provided a fairly complete listing of woody plant sensitivities to O_3 exposure.

Ozone injury in broadleaved species is typically characterized by necrotic spots (stipples or flecks) on foliar surfaces. Mottling, distal necrosis, browning, premature needle shed, or a general decline typically occurs in needled species such as the pines. While all of the physiological effects of O_3 injury are not adequately defined, palisade cells appear to be most directly affected by O_3 exposure (Smith, 1981).

From the available literature and air quality monitoring information, it appears that over widespread regions in the eastern United States episodes of elevated O_3 concentrations occur that are above threshold levels thought to be harmful to trees. Even relatively remote areas such as Acadia National Park in Maine have recently been shown to exhibit evidence of O_3 damage to eastern white pine. The picture which emerges on the O_3 issue is that much of our rural forestland may be exposed to episodic, and possibly chronic, O_3 concentrations which have the potential to affect forest health either directly or through interactions with other pollutant and natural stress factors.

(2) Fertilization Effect

The emission of materials by anthropogenic activities with subsequent deposition of these materials to forested ecosystems is a process involving most of the 17 essential nutrients for plant growth. Indeed the elements primarily responsible for acid deposition are N and S, both of which are frequently applied to crops in the form of fertilizers to improve growth. In forestlands, nitrogen is the most commonly applied fertilizer nutrient since it is the most commonly limiting nutrient in forested ecosystems.

Nitrogen deposition has become a focus of interest due to the possibility of a fertilization effect on forests from atmospheric deposition. Uptake of atmospherically derived N could be directly through foliar organs or indirectly via the soil. Lovett et al. (1982) estimated wet deposition of N in a high elevation site in New Hampshire to be 44.1 kg N ha^{-1} yr^{-1}. An estimate of wet deposition for low elevation sites in the northeast would range between 3 and 18 kg N ha^{-1} yr^{-1} depending on the location (NADP, 1985). These estimates do not include dry deposition contributions of N which may be significant. At this rate of deposition to forests typically deficient in N, we could expect a stimulation of tree growth to result. This in itself would be a welcome contribution to forest production. However, other consequences may result from additions of N to the system which may have detrimental effects on trees.

Schutt and Cowling (1985) recently grouped the major concerns for atmospheric deposition of N to forests into six types of effects.

(a) Nitrogen may increase tree growth rates resulting in greater demands for other essential elements leading to deficiencies (Abrahamson, 1980) or imbalances (Friedland et al. 1985) in nutrient supply.

(b) Nitrogen may inhibit infection or induce necrosis of mycorrhizae, a possibility supported by Shafer et al. (1985) in recent experiments with simulated acidic rain.

(c) Nitrogen may increase the susceptibility of trees to winter damage as hypothesized by Friedland et al. (1984b) to explain the decline of high elevation red spruce in the northeast. Reduced hardiness to cold temperatures could result from delayed cuticularization and conversion of starch to sugars.

(d) Nitrogen may increase susceptibility to root-disease fungi.

(e) Nitrogen may alter root-shoot ratios.

(f) Nitrogen deposition may alter patterns of nitrification, denitrification, and possibly nitrogen fixation.

(3) Acid Deposition

The potential for acid deposition to alter the health of trees could result from direct effects of wet and dry S and N deposition on foliage, or from indirect effects due to chemical alterations in the soil. Little evidence exists to suggest that current average ambient levels of precipitation acidity would cause lesions, necrosis, or cuticular degradation of foliar surfaces. Morrison (1984) reviewed the literature dealing with acid deposition effects on forests and found the threshold for injury to foliage in experiments utilizing simulated acid rain ranged in pH from 2.0 to 3.0. The annual precipitation weighted mean hydrogen ion concentrations expressed as pH reported for the United States in 1982 were all above a pH of 4.0 (NADP, 1985), which is one to two orders of magnitude less acidic than the thresholds for foliar damage summarized by Morrison. Tree foliage may, however, be subjected to solutions of lower pH than 4.0 by interception of low pH cloud droplets, during low pH precipitation events, by evaporation and subsequent concentration of solutions on foliar surfaces, by dissolution of acidic particulates dry deposited on foliage prior to a precipitation event, or possibly by throughfall which has been acidified in the upper canopy. Very little research is available to evaluate the extent of these processes.

One of the symptoms associated with West Germany's "Waldsterben" is the yellowing (i.e. chlorosis) of Norway spruce foliage at high elevations. The group of scientists led by Professor Rehfuess at the University of Munich believe this may be the result of acid deposition induced leaching of nutrients from the foliage and soil, resulting in deficiencies of nutrients and the symptomatic chlorosis (Rehfuess, 1983; Cowling, 1984). From their investigations it appears that primarily magnesium (Mg) and possibly calcium (Ca) are the nutrient elements involved in the foliar leaching effect.

Numerous studies have examined the potential indirect effects of acid deposition on forest productivity through soil related mechanisms. This body of literature is discussed by Bockheim (1984), Fernandez (1985), Morrison (1984) and U.S. EPA (1983a).

An obvious initial concern is for the possibility of acid deposition causing an acidification of the soil with subsequent harmful effects on tree growth. Along with soil acidification we could expect a possible decline in soil base saturation leading to a reduced availability of the major base cationic nutrient elements (e.g. Ca^{2+}, Mg^{2+}, K^+). The potential accelerated leaching of basic cations may be more directly related to mobile anions in soil solutions, such as SO_4^{2-}, than to H^+ itself. While these effects could occur over long periods of time, there is no evidence that forest soils have been significantly acidified in the U.S. by acid deposition. Some of the reasons for this current lack of evidence on soil acidification are as follows:

(a) Forest soils are typically highly acidic due to natural soil forming processes. The average soil pH from 25 forested watersheds sampled in Maine in 1984 was pH = 3.29 (O horizon), pH = 3.48 (E horizon), pH = 4.24 (B horizon), and pH = 4.60 (C horizon).

(b) The natural processes which occur in forest soils generate a large amount of acidity, and atmospherically derived acidity is typically a small increment to the total production of acidity in the soil.

(c) The weathering of soil parent materials consumes H^+ and releases basic cations. This is true even for the so called "acid forming" rock types such as the aluminosilicate group. The rate of mineral weathering is often difficult to quantify for a particular site. If mineral weathering rates are sufficient to resupply soil cation pools, no acidification due to acid deposition would be expected. Clearly the contribution of mineral weathering to soil buffering needs to be better quantified.

This is not to say that the deposition of atmospheric H_2SO_4 and HNO_3 has no effect on forest soils. There is no question that modern inputs of strong mineral acids have changed the chemical character of soils and soil solutions, but understanding the implications of these changes on forest growth will be a complex task. For example, Fernandez and Struchtemeyer (1985) found that the balance between Ca^{2+} and Al^{3+} in the forest floor of spruce-fir stands in Maine appeared to be related to site productivity. If acid deposition alters the balance between critical elements in the soil, it seems possible that subtle changes in the rate of tree growth could occur without the wholesale forest deterioration and soil acidification often expected as evidence of an effect. These subtle effects, however, are difficult to define within the complexity of forested ecosystems. It should also be noted that evidence of soil acidification by acid deposition appears more likely in central Europe, where current rates of total deposition are higher and there exists a longer history of air pollution.

One of the key concerns regarding the biological effects of atmospheric deposition on soils has been the mobilization of Al^{3+} into soil solutions. Aluminum is naturally abundant in soils, but Al is often immobile due to the low solubility of Al containing minerals. Acid deposition may increase the mobility of soil Al by (a) exchange between atmospherically derived H^+ and adsorbed Al^{3+} on soil colloid exchange sites or (b) dissolution of Al from soil minerals which typically results in the consumption of H^+. Ulrich et al. (1980) and Ulrich (1983) present the hypothesis that the net effect of acid deposition to soils in the Solling region of West Germany is the mobilization of soil Al which has toxic effects on tree roots and contributes to the decline of the forest. Although this hypothesis may be valid for the Solling, no evidence to date supports this as the mechanism responsible for forest deterioration in the eastern United States.

(4) Trace Metals

There is currently no continuous large scale network for monitoring the deposition of trace metals (e.g. Pb, Cd, Cu, Ni, Zn) in rural forested regions as exists for monitoring the wet deposition of acids and major ions within the National Atmospheric Deposition Program (NADP). However, from the measurements which have been made it appears that forests almost everywhere are receiving inputs of trace metals from the atmosphere well above what would be expected in the absence of air pollution. Deposition of trace metals is generally associated with particulates close to the sources of emission, with a greater proportion of trace metal

inputs being associated with wet deposition farther from the
sources. The literature does show, for example, that the
metropolitan corridor in the northeast (which includes Washington
D.C., New York, and Boston) is a region of relatively high
concentrations of metals in the atmosphere such as Pb (U.S. EPA,
1983b). Peters and Bonelli (1982) presented Pb loading rates from
bulk precipitation with filtered samples for the north-central and
northeastern regions. While these data are significantly lower
than what has been reported in the literature for unfiltered
samples (presumably due to the trace metals associated with
particulates), they do demonstrate the breadth of geographic area
subjected to elevated levels of Pb deposition.

We can view the subject of trace metal deposition to forests
as four categories of information needs.

      (a) Deposition Rates

      (b) Accumulation in the Ecosystem

      (c) Mobilization and Speciation of Metals

      (d) Toxicity to Biota

(a)   As discussed above, little continuous monitoring information
is available for trace metal deposition in rural forested
areas. We do have an understanding of the complexity of dry
deposition to forest canopies, as discussed by Lovett (1984),
but available monitoring technology for dry deposition is
typically expensive which inhibits the development of
extensive monitoring systems. The U.S. Environmental
Protection Agency currently has plans for developing a network
for dry deposition monitoring at up to 100 sites which should
greatly improve the available data on trace metal inputs.
However, these will not primarily be sites under forest
canopies which can have a major effect on total trace metal
deposition to the site due to interception of particulates by
the canopy.

(b)   The accumulation of trace metals in forested ecosystems is
well documented. Because of the ability of trace metals to be
complexed by organic matter, much of the accumulation in
forested ecosystems occurs in the forest floor. The
occurrence of accumulated trace metals in the forest floor in

the northeastern U.S. has been well documented (Johnson et al. 1982, Andresen et al. 1980, Friedland et al. 1984a). Hanson et al. (1982) showed that the trend for trace metal concentrations in the forest floor at high elevation sites was to decrease along a southwest to northeast axis with distance from the northeast metropolitan corridor. Fernandez and Czapowskyj (1985) reported that even low elevation, commercial forest sites in northern Maine contained trace metal concentrations that appeared to be above what would be expected for background conditions. Therefore, little doubt exists that atmospherically derived trace metals are accumulating in forested ecosystems, but the residence time, mobilization, speciation, and effects of these metals are poorly defined.

(c) A key question with regard to forest soil accumulations of trace metals relates to the conditions and rates of mobilization for trace metals in soil. A great deal of literature is available on trace metal mobility in soils from research on mine reclamation, land applications of effluent and sludge, and basic soil chemistry research under agronomic conditions (Elliott and Stevenson, 1977; Pendias and Kabata-Pendias, 1984). Very little work has been done to determine the rates of trace metal mobilization from the forest floor and the migration of these metals through soil profiles. The speciation of trace metals in forest soil solutions under field conditions is also a critical information need. Trace metals can occur in various ionic states in soil solutions as well as in complexes with organic acids which significantly affects the toxicity of these metals to living organisms.

(d) Ultimately we must understand the toxicity and dose-response relationships between forest soil and soil solution trace metal concentrations and effects on soil microorganisms, trees, and wildlife. The role an individual metal plays in the health of forests may very well be obscured by interactive and synergistic effects among that metal and other trace

metals, other air pollutants, and natural stress factors in
the environment. Of the possible effects of trace metal
accumulations in the forest floor, one which has been studied
is the inhibition of organic matter decomposition (Tyler 1975,
Ruhling and Tyler 1973, Moloney et al. 1983). Although
laboratory and manipulative studies have shown the possibility
of a suppression of soil microbiological activity, no studies
to date have clearly demonstrated that this is occurring under
field situations. Only a small portion of the pool of
accumulated trace metals in forest floors appears to be
assimilated by trees, and little is known at present about the
consequences of ambient soil solution trace metal
concentrations to nutrient uptake or to the physiological
functioning of the tree.

(5) General Stress

The "General Stress" and "Growth Altering Substances"
hypotheses were developed by West German scientists and have been
recently summarized by Schutt and Cowling (1985). Essentially the
General Stress hypothesis suggests that atmospheric deposition has
caused a decrease in net photosynthesis and a diversion of
photosynthate from mobile carbohydrates to less mobile secondary
metabolites. This results in a lower energy status in the tree and
increased susceptibility to other stress factors such as insects,
disease, nutrient deficiencies, drought, or frost damage.

(6) Growth Altering Substances

The Growth Altering Substances hypothesis deals with the
possibility that some of the numerous synthetic organic compounds
produced each year may be toxic to trees if emitted into the
atmosphere and deposited in forested ecosystems (Schutt and
Cowling, 1985). Almost no monitoring data exist for most of the
chemicals possibly involved, and no case of a regional forest
effect from growth altering organic substances in the U.S. is
evident in the scientific literature. The absence of information,
however, does not exclude this mechanism of forest effect from
being worthy of investigation as a possible contributor to modern
forest deterioration phenomena.

11.5 A GENERIC HYPOTHESIS FOR FOREST DETERIORATION

It is evident from this discussion that the suite of modern air
pollutants deposited in forested ecosystems may have numerous effects
on the health of forests. Given the current evidence, it does not seem
reasonable to anticipate that a single mechanism of forest effect will
be identified which can explain the evidence of present or future
forest deterioration at all locations, assuming air pollutants are
involved at all. For example, we might suspect that acid cloud
moisture and O_3 have a higher probability of playing a role in forest
health at high elevation southern Appalachian sites, whereas more
subtle effects from soil leaching and trace metal accumulations may be
critical to productivity in commercial forests of northern New
England. To incorporate the evidence available into a single
framework, we might view the possible role of air pollution in forest
deterioration as a generic hypothesis in three stages.

(1) Primary Chronic Stress Factors

Primary chronic stress factors may induce stress in forests
over long periods of time, essentially predisposing the tree to
negative effects from subsequent stress factors in the
environment. This category could include one or more factors such
as harvesting, trends in climate, stand dynamics, and a wide range
of air pollutants. No visually observable effects on tree health
would be associated with these factors, but alterations in
biochemical pathways and the energy status of the tree are possible
as suggested in the West German General Stress hypothesis. Trees
would presumably recover if the operative stress factors in this
group were removed or ameliorated.

(2) Primary Acute Stress Factors

Trees weakened by chronic stress are more susceptible to
subsequent acute stress which could then be responsible for
initiating a deterioration syndrome of the forest. Acute stress
factors could include a period of elevated air pollutant deposition
(that is thought to have occured since the 1950's), periods of
drought, cold temperatures, or intensive harvesting. Once the
deterioration has begun in response to acute stress, recovery of
the affected trees seems doubtful even when the acute stress

factors are removed. This does not mean, however, that net forest productivity or the vigor of regeneration would necessarily be diminished.

(3) Secondary Stress Factors

Once primary stress factors have induced a forest deterioration, other stresses participate in adversely affecting the forest. Frequently, secondary stress would result from insects and disease which characteristically attack already weakened individuals. Mechanical stress and changes in soil temperature and moisture from a gradual thinning of the canopy due to foliage loss and mortality may enhance the trend in deterioration within the stand.

This framework permits numerous viable mechanisms in light of the diversity of symptoms, species, soil types, climatic regimes, and deposition rates which are implicated in this issue.

11.6 WHAT DO WE KNOW TO DATE?

To determine the role that atmospheric deposition plays in forest management decision making processes, it is useful to summarize specific strengths and weaknesses in our understanding of this issue. Several key points are identifiable.

(1) We do not know

(a) the mechanisms of atmospheric deposition effects on forests responsible, if any, for the modern deterioration of forests under a variety of environmental conditions.

(b) the deposition of gaseous pollutants, trace metals, particulates, cloud moisture, and organic compounds to most remote forested areas. We do have in place a good network for precipitation monitoring, with measurements of major cations and anions, but little monitoring information is available on throughfall characteristics after precipitation interacts with forest canopies.

(c) the dose-response relationships for most pollutants under ambient field conditions in forested ecosystems. More importantly, we have a very poor understanding of the effects on forests of interactions between individual pollutants, and between pollutants and natural stresses.

(d) the chronic and the acute natural and air pollutant stress factors responsible for the unexplained regional forest deterioration.

(2) We do know

(a) that forested ecosystems almost everywhere are exposed to a modern chemical climate unique in the evolutionary history of these vegetative communities.

(b) that in the eastern U.S., a number of high elevation coniferous sites have exhibited an unexplained deterioration of the forest with red spruce showing the greatest effect. Other forest species and types appear to also be involved but drawing definitive conclusions remains difficult. Several species are involved in the central European Waldsterben, with the extent of deterioration apparently much more advanced than in the U.S. at the present time.

(c) that soils and soil solutions in forests exhibit altered chemical characteristics due to atmospheric deposition. Forest floors have accumulated trace metals; soil solutions are dominated by the strong mineral acid anion SO_4^{2-}; and sites receiving high rates of deposition show elevated N in the system.

11.7 OTHER CONSIDERATIONS

It is particularly important for those investigating atmospheric deposition effects on forests, as well as for the public and policy makers, to maintain a broad perspective on the complexity of factors which may influence forest health. This discussion has focused on O_3, S, N, and trace metals relative to forest effects, but other factors must be simultaneously considered. For example, according to paleoecological records whole populations of tree species have been replaced by other species, only to return to dominance at a later time. Why? Also, forest harvesting in modern, cultured forestlands is frequent and biomass removals are intensifying. What role does this vehicle for nutrient export from the site play in stand vigor when compared to acid deposition induced nutrient leaching? What pests or pathogens might be involved in forest deterioration which we may not

yet have identified? What other air pollutants, such as fluorides or
other photochemical oxidants, may be important? When we speak of the
effects of air pollutants on forests, we seldom include CO_2 in these
discussions, yet our atmosphere is becoming enriched with CO_2 and the
controversy continues on how climate may change as a result (Idso,
1982). Since we can "fertilize" trees with CO_2, and since it is
possible that precipitation and temperature trends have or will change
independently or as a result of CO_2 increases, then should
investigations of multiple air pollution stresses include a CO_2
component? These are only a few of the unknowns suggesting the need
for atmospheric deposition effects to be evaluated within a broad
perspective relative to the complexity of the environment which
influences forests.

11.8 CONCLUSIONS

The practical question which emerges is whether regional air
pollution has become a factor in forest stress management. It does not
appear reasonable, at this time, to recommend changes in stand
prescriptions based on the available evidence for forest effects from
regional air pollution. Without a well defined mechanism of stress at
a particular site, clearly no adjustments in management decisions could
appropriately be made. Depending on the role air pollution might
eventually be shown to play in forest health, there are options
available to be considered by forest managers in addressing the
problem. These options could affect decisions on species selection,
fertilization, liming, site preparation, harvesting schedules and
intensities, and pesticide programs, as well as decisions on the sale
and acquisition of land. Obviously a program of emission reduction for
the pollutants of concern would be a useful long-term approach to
alleviating identified air pollutant effect problems.

It is incumbent upon all forestry professionals to keep informed on
the atmospheric deposition effects issue. We are all exposed to a wide
range of perspectives on this topic in the popular press and scientific
literature, including a diverse menu of legislation proposed to address
the problem. Foresters will be called upon for their insight on this
issue, and need to be informed to promote the most appropriate actions
in management and policy decisions.

ACKNOWLEDGEMENTS

The author wishes to express his appreciation to Mary Thibodeau, Jeffrey Risser, and Larry Zibilske for their comments and assistance in the preparation of this manuscript.

REFERENCES

1. Abrahamson, G. 1980. Acid precipitation, plant nutrients, and forest growth. Proc. Int. Conf. Ecol. Impacts Acid Precipitation. SNSF Project, As, Norway.
2. Andresen, Anthony M., Arthur H. Johnson, and Thomas Siccama. 1980. Levels of lead, copper, and zinc in the forest floor in the northeastern United States. J. Environ. Qual. 9:293-296.
3. Bernsten, C.M., D.W. Johnson, J.F. Corliss, H.C. Jones, I.J. Fernandez, and W.H. Smith. 1984. Acidic deposition and forests. Report of the SAF Task Force on the Effects of Acidic Deposition on Forest Ecosystems. Society of American Foresters. 51pp.
4. Bockheim, J.G. 1984. Acidic deposition effects on forest soils and site quality. In: Proc. U.S. - Canadian Conference on Forest Responses to Acidic Deposition. Land and Water Resources Center, University of Maine, Orono, Maine. pp. 19-35.
5. Breece, Linda and Sherman Hasbrouck (eds.). 1984. Forest responses to acidic deposition. Proc: U.S. - Canadian Conf. on Forest Responses to Acidic Deposition. Land and Water Resources Center, University of Maine, Orono, Maine.
6. Burgess, Robert L. (ed.). 1984. Effects of acidic deposition on forest ecosystems in the northeastern United States: An evaluation of current evidence. ESF 84-016. College of Env. Sci. and For., Syracuse, NY. 140pp.
7. Chevone, Boris I. 1985. Gaseous and wet atmospheric pollutant effects on forest tree seedling growth. In: Proc. Symposium on Air Pollutants Effects on Foret Ecosystems. The Acid Rain Foundation, St. Paul, Minnesota. pp. 87-94.
8. Council for Agricultural Science and Technology (CAST). 1984. Acid precipitation in relation to agriculture, forestry, and aquatic biology. C.A.S.T. Report No. 100. Ames, Iowa. 31pp.
9. Cowling, Ellis B. 1984. Conclusions regarding the decline of forests in North America and central Europe. Statement prepared for William Ruckelshaus, Administrator, U.S. Environmental Protection Agency. North Carolina State University, Raleigh, North Carolina. 13pp.
10. Davis, D.D. and R.G. Wilhour. 1976. Susceptibility of woody plants to sulfur dioxide and photochemical oxidants. U.S. Environmental Protection Agency Ecol. Res. Series EPA-600/3-76-102. Corvallis, Oregon. 72pp.
11. Elliott, L.F. and F.J. Stevenson (eds.). 1977. Soils for management of organic wastes and waste waters. American Society of Agronomy, Madison, Wis. 650 pp.

12. Fernandez, Ivan J. 1983. Acidic deposition and its effects on forest productivity - A review of the present state of knowledge, research activities, and information needs. National Council of the Paper Industry for Air and Stream Improvement Tech. Bull. 392. New York. 104pp.

13. Fernandez, Ivan J. 1985. Potential effects of atmospheric deposition on forest soils. In: Proc. Symposium on Air Pollutants Effects on Forest Ecosystems. The Acid Rain Foundation, St. Paul, Minnesota. pp. 238-250.

14. Fernandez, I.J. and R.A. Struchtemeyer. 1985. Chemical characteristics of soils under spruce-fir forests in eastern Maine. Can. J. Soil Sci. 65:61-69.

15. Fernandez, Ivan J. and Miroslaw M. Czapowskyj. 1985. Levels of trace metals in the forest floors of low elevation, commercial spruce-fir sites in Maine. Northeast. Env. Sci. 4:1-7.

16. Friedland, A.J., A.H. Johnson, T.G. Siccama, and D.L. Mader. 1984a. Trace metal profiles in the forest floor of New England. Soil Sci. Soc. Am. J. 48:422-425.

17. Friedland, Andrew J., Gary J. Hawley, and Robert A. Gregory. 1985. Investigations of nitrogen as a possible contributor to red spruce (Picea rubens Sarg.) decline. In: Proc. Symposium on Air Pollutants Effects on Forest Ecosystems. The Acid Rain Foundation, St. Paul, Minnesota. pp. 95-106.

18. Friedland, Andrew J., Robert A. Gregory, Lauri Karenlampi, and Arthur H. Johnson. 1984b. Winter damage to foliage as a factor in red spruce decline. Can. J. For. Res. 14:963-965.

19. Hanson, Denis W., Stephen A. Norton, and John S. Williams. 1982. Modern and paleolimnological evidence for accelerated leaching and metal accumulation in soils in New England caused by atmospheric deposition. Water Air Soil Pollut. 18:227-239.

20. Idso, Sherwood B. 1982. Carbon dioxide: Friend or foe? IBR Press, Tempe, Arizona. 92pp.

21. Johnson, A.H. and T.G. Siccama. 1983. Acid deposition and forest decline. Environ. Sci. Tech. 17:294-306.

22. Johnson, A.H., T.G. Siccama, and A.J. Friedland. 1982. Spatial and temporal patterns of lead accumulation in the forest floor in the northeastern United States. J. Environ. Qual. 11:577-580.

23. Lovett, Gary M. 1984. Atmospheric deposition to forests. In: Proc. U.S. - Canadian Conference on Forest Responses to Acidic Deposition. Land and Water Resources Center, University of Maine, Orono, Maine. pp. 7-18.

24. Lovett, Gary M., William A. Reiners, and Richard K. Olsen. 1982. Cloud droplet deposition in subalpine balsam fir forests: Hydrological and chemical inputs. Science 218:1303-1304.

25. Miller, P.R., O.C. Taylor, and R.G. Wilhour. 1982. Oxidant air pollution effects on a western coniferous forest ecosystem. U.S. Environmental Protection Agency Public. No. EPA-600/D-82-276, Corvallis, Oregon. 10pp.

26. Moloney, K.A., L.J. Stratton, and R.M. Klein. 1983. Effects of simulated acidic, metal-containing precipitation on coniferous litter decomposition. Can. J. Bot. 61:3337-3342.

27. Morrison, Ian K. 1984. Acid rain - A review of literature on acid deposition effects in forest ecosystems. For. Abst. 45:483-506.

28. National Atmospheric Deposition Program. 1985. NADP annual data summary - Precipitation chemistry in the United States - 1982. Colorado State University, Fort Collins, Colorado.
29. National Swedish Environment Protection Board (NSEPB). 1982. Ecological effects of acid deposition. Report on background papers. Stockholm Conference on the Acidification of the Environment. Expert Meeting I. Report SNV PM 1636. Stockholm, Sweden.
30. Pendias, Henryk and Alina Kabata-Pendias. 1984. Trace metals in soils and plants. CRC Press, Inc., Boca Raton, Florida. 315 pp.
31. Peters, Norman E. and Joseph E. Bonelli. 1982. Chemical composition of bulk precipitation in the north-central and northeastern United States, December 1980 through February 1981. U.S. Geological Survey Circular 874. Alexandria, VA.
32. Rehfuess, K.E. 1983. Walderkrankugen und Immissionen-eine Zwischenbilanz. Allg. Forstzeitschr. 38:601-610.
33. Ruhling, Ake and Germund Tyler. 1973. Heavy metal pollution and decomposition of spruce needle litter. Oikos 24:402-41630.
34. Schutt, Peter and Ellis B. Cowling. 1985. Waldesterben, a general decline of forests in central Europe: Symptoms, development, and possible causes. Plant Disease 69:548-558.
35. Shafer, S.R., R.I. Bruck, and A.S. Heagle. 1985. Formation of ectomycorrhizae on Pinus taeda seedlings exposed to simulated acidic rain. Can. J. For. Res. 15:66-71.
36. Skelly, John M. 1980. Photochemical oxidant impact on Mediterranean and temperate forest ecosystems: Real and potential effects. In: Proc. Symposium on Effects of Air Pollutants on Mediterranean and Temperate Forest Ecosystems. Pacific Southwest For. and Range Experiment Station, Berkley, California. pp. 38-50.
37. Smith, W.H. 1981. Air pollution and forests. Springer-Verlag, Inc. New York, NY. 379pp.
38. The Acid Rain Foundation (TARF). 1985. Proc.: Symposium on Air pollutants effects on forest ecosystems. St. Paul, Minnesota.
39. Tyler, Germund. 1975. Effect of heavy metal pollution on decomposition and mineralization rates in forest soils. In: International Conference on Heavy Metals in the Environment. Toronto, Canada. pp.217-226.
40. Ulrich, B. 1983. A concept of forest ecosystem stability and of acid deposition as driving force for destabilization. In: Effects of Accumulation of Air Pollutants in Forest Ecosystems. D. Reidel Pub. Co., Boston, MA. pp.1-32.
41. Ulrich, B., R. Mayer and P.K. Khanna. 1980. Chemical changes due to acid precipitation in a loess-derived soil in central Europe. Soil Sci. 130:193-199.
42. U.S. Environmental Protection Agency. 1983a. The acidic deposition phenomenon and its effects. Critical Assessment Review Papers, Vol. II. Effects Sciences. EPA-600/8-83-0116B. Washington, D.C.
43. U.S. Environmental Protection Agency. 1983b. National air quality and emissions trends report, 1983. EPA-450/4-84-029. Washington, D.C.
44. United States Forest Service. 1985. Effects of atmospheric deposition on spruce-fir forests: Research plan for a spruce-fir effects cooperative. U.S. Forest Service Northeastern For. Experiment Station, Broomall, PA.

45. von Osten, Wolf U. 1985. Evidence for effects of air pollution on ecosystems, in particular on forests in the Federal Republic of Germany. In: Proc. Symposium on Air Pollutants Effects on Forest Ecosystems. The Acid Rain Foundation, St. Paul, Minnesota. pp.175-190.